THE GREAT VICTORIA DESERT

The fat-tailed dunnart
Sminthopsis crassicaudata

THE GREAT VICTORIA DESERT

North of the Nullarbor
South of the Centre

MARK SHEPHARD

REED

First published in 1995 by
REED BOOKS
a part of Reed Books Australia
Level 9, North Tower
1–5 Railway Street
Chatswood, NSW 2067

National Library of Australia
Cataloguing-in-Publication Data

Shephard, Mark.
The Great Victoria Desert

Bibliography.
Includes index.
ISBN 0 7301 0485 0.

1. Desert fauna - Great Victoria Desert (S. Aust and W.A.).
2. Desert biology - Great Victoria Desert (S. Aust and W.A.).
3. Conservation of natural resources - Great Victoria Desert (S. Aust and W.A.).
4. Great Victoria Desert (S. Aust and W.A.) - History.
5. Great Victoria Desert (S. Aust and W.A.). I. Title.

574.52652099415

Edited by Sharon Paull
Designed by Trevor Hood/ANACONDA GRAPHIC DESIGN
Maps by David Hayes/LOFTY DESIGNS

Printed in Hong Kong by South China Printing

Contents

This book is intended to represent a personal appreciation of the Great Victoria Desert; my contribution to the dissemination of information about a little known, but totally inspiring part of Australia's natural and cultural heritage. The desert has a fascinating, yet controversial and emotive history. Many sensitive issues, such as atomic testing on Australian soil, religious teachings at Christian-based missions around the desert, and Aboriginal people's struggle for land rights and a return to traditional lands, have been addressed as part of the desert's 'story'. I have attempted to present the facts relating to these issues in a clear and concise manner. I trust the book will be regarded as a positive and constructive contribution to popular literature on a magnificent area of Australian wilderness.

I have received a great deal of help in writing this book and I am particularly keen to acknowledge the contribution of these helpers. The following people have reviewed individual chapters:

Introduction	Ross Allen, Andy Chapman
Chapter 1	Mark Benbow, Peter Copley, Graham Krieg, Bob Major
Chapter 2	Peter Canty, Andy Chapman, Josephine Morelli
Chapter 3	Dr Andrew Black, Graham Carpenter, Andy Chapman, Peter Copley
Chapter 4	Andy Chapman, Peter Copley, Dr Catherine Kemper
Chapter 5	Andy Chapman, Peter Copley, Dr Mark Hutchinson
Chapter 6	Tom Gara, Ushma Scales, Jan Turner, Bob Ware
Chapter 7	Tom Gara, Bruce Macdonald
Chapter 8	Bill Edwards, Tom Gara, Ushma Scales, Dr John Stanton, Jan Turner, Bob Ware
Chapter 9	Len Beadell, Bob Major, Ushma Scales, Jan Turner, Bob Ware
Chapter 10	Len and Anne Beadell, Ushma Scales, Jan Turner
Chapter 11	Jan Turner
Chapter 12	Ross Allen, Andy Chapman, Jan Turner, Ushma Scales, Bob Ware

I wish to make particular mention of Andy Chapman (CALM, Kalgoorlie), David Pearson (CALM, Perth) and Jan Turner (staff anthropologist, Ngaanyatjarra Council) who, although many hundreds of kilometres away, patiently and promptly answered my numerous telephone calls and faxes for help and information.

A number of other people have contributed material for, or reviewed, particular sections of the book. These include:

Brenton Arnold (Tallaringa Conservation Park)
Ian Baird (Cundeelee Mission and Tjuntjuntjara)
Archie Barton (Maralinga Tjarutja)
John Blockley, Bill Carr, Vera Novak, Pietra Guj (mining)
Nell Charlwood (East Brothers)
Mike Donaldson (Frank Hann and other explorers)
Wilf Douglas (missions)
Mary Durack (Aboriginal land tenure)
Fern Hames and Sandra Boyd (Tallaringa Conservation Park)
Brian Hadfield (Cundeelee Mission)
George Kwitko (mining)
John Morris (Maralinga)
Pia Richter (Unnamed Conservation Park)
Ronald Schenk (Mt Margaret and the Warburton Ranges Mission)

The following people willingly donated their photographs for inclusion in the book: Ruth Aspley, Len Beadell, Mark Benbow, Archie Barton on behalf of Maralinga Tjarutja, Peter Canty, Andy Chapman, Simon Coxon, Mike Donaldson, Lynton Huxley, Beryl Mazzachi, Heinz and Chloe Moeller, David Pearson, Lyn Pedler, John Streich, Chris Wenham and George Wood. Photographs of members of the Irrunytju (Wingellina) Community were taken by Anne and Ken Scott and are reproduced with the permission of the Community and the Ngaanyatjarra Council. I am very appreciative of the Irrunytju Community's willingness to assist this project. Photographs of Tallaringa Conservation Park were taken by Dr Kathy Dynes and kindly supplied by the Australian and New Zealand Scientific Expedition Society (ANZSES). Selected historical black and white photographs were obtained from the Mortlock Library of South Australiana and the South Australian Branch of the Royal Geographical Society of Australasia. Neville Gilbourne, John Clarke and Norma Knopp (Duckpond Pty Ltd) are thanked for their continued friendly service in developing and duplicating the dozens of rolls of photographic film taken by me in the desert.

There are several friends who have accompanied me on field trips to the Great Victoria Desert. I would like to thank them for their support and encouragement throughout this project. George and Ann Wood have become life-long friends as a result of a mutual interest in and fascination with the desert. George and Ruth Aspley, who took me on my first desert experience in the mid-1980s, have remained wonderful friends and travelling companions. Len and Anne Beadell inspired me to write this book, and I was privileged to be part of a private trip to the desert with the Beadells and their daughter Connie in 1991.

Karen and Lynton Huxley, Lawrie and Meryn Money, Josephine Pyle, Emanuel Saris, and Michael and Karen Wood have also been part of desert trips with me and my family. John Jeffries (Sokkia Pty Ltd, Keswick, South Australia) is thanked for providing me with the use of a GPS system for some of my field work in the desert.

I am particularly grateful for the help given to me by members of the Ngaanyatjarra Council, Pitjantjatjara Council and Maralinga Tjarutja. I have been very fortunate to meet and work with many of the Aboriginal people from communities in and around the Great Victoria Desert over the past three years.

Lesley Hutton (Southern Wordprocessing Service) meticulously typed the many drafts of the manuscript. I am privileged to have been able to use her exceptional typing talents over many years. David Hayes (Lofty Designs, Longwood, SA) prepared the maps for the book. Helen Strafford compiled the index.

Finally, I would like to dedicate this book to my wife Anne, son Matthew and daughter Emily, who have so patiently supported me throughout the many long hours which it has taken to compile the book.

SETTING THE RECORD STRAIGHT

If ten Australians were asked the question 'Where is the Great Victoria Desert?', a variety of different responses would most likely be given. Many would have no perception at all as to the desert's location, while some would place the desert by association somewhere in the State of Victoria.

Those who know the Great Victoria Desert would argue that a general lack of knowledge of the desert by the public is good — while it remains unknown, this hidden jewel of Australia's outback is safe from environmental degradation and human misuse, whether accidental or unintentional. For it is the desert's integrity, its very real wilderness quality, which makes it such a special place.

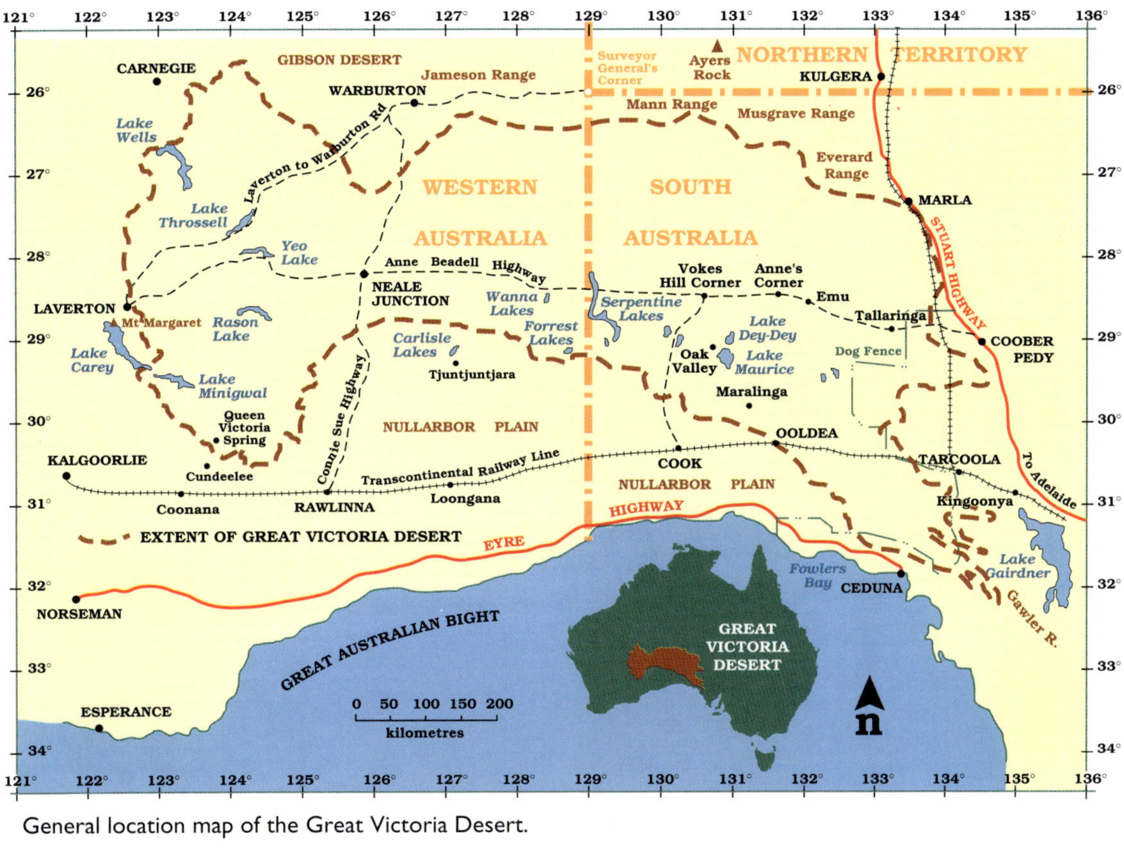

General location map of the Great Victoria Desert.

Mark Benbow, Mines and Energy SA

ABOVE: Parallel dunefields, which are aligned in a west–east direction, are a dominant landform of the Great Victoria Desert.

BELOW: The Great Victoria Desert and Queen Victoria Spring (pictured) were named by Ernest Giles in September 1875.

David Pearson

RIGHT AND BELOW: The soft sands of the Great Victoria Desert are shaped into spectacular dunes by prevailing westerly winds.

Mark Shephard

George Wood

Mark Shephard

The marble gum Eucalyptus gongylocarpa *is the 'signature' tree of the Great Victoria Desert.*

Others would suggest that the unique attributes of one of Australia's greatest natural and cultural regions should be more widely promoted and that the area should emerge from behind its hidden veil.

Where Is the Great Victoria Desert?

The boundaries of the Great Victoria Desert straddle two States, Western Australia and South Australia.[1,2] To the south the desert merges with the northern edge of the Nullarbor Plain and the nearby Transcontinental (or Trans-Australian) Railway Line which links Adelaide and Perth. The ironstone gravel plains of the Gibson Desert and a series of ranges stretching from Warburton in Western Australia to the Anangu Pitjantjatjara Lands in South Australia form the northern boundary of the desert. These ranges include the Warburton, Blackstone, Jameson, Mann, Tomkinson, Birksgate, Musgrave and Everard Ranges. Mabel Creek pastoral station, adjacent to the town of Coober Pedy on the Stuart Highway, represents the desert's eastern limit, while Laverton in Western Australia is centred on the desert's western boundary. The total area occupied by the desert is approximately 42 million hectares (420,000 square kilometres).

What Is the Great Victoria Desert?

The Great Victoria Desert is a vast sand belt comprising both sandplains and sand dunes.[1] The dunefields are long and linear, and run parallel to each other in a general west-to-east direction; this alignment is in contrast to the more well-known Simpson Desert dunefields which run essentially north to south.

Mark Shephard

Peter Canty

ABOVE: Aboriginal people have lived in the Great Victoria Desert for over 20,000 years. Evidence of their traditional occupation is scattered widely throughout the desert.

RIGHT: Field studies during the past 25 years have shown that the Great Victoria Desert supports an extraordinary diversity of reptiles. The pale knob-tailed gecko Nephrurus laevissimus *is a widespread inhabitant of the desert.*

Why 'Great Victoria'?

The Great Victoria Desert was named after Queen Victoria by the explorer Ernest Giles in 1875. Giles had undertaken a 17-day, 500-kilometre, waterless journey across the desert's heart. On reaching its south-western edge, he arrived at a small claypan brimming with water. He camped at the claypan, which he named Queen Victoria Spring, and remarked in his diary 'the Great Victoria Desert...which will probably extend to the west as far as it does to the east, I have also honoured with her Majesty's mighty name, calling it the Great Victoria Desert'.[3]

Why 'Desert'?

Australian 'deserts', including the Great Victoria Desert, generally show the following characteristics:

- they are areas largely uninhabited by humans;
- they receive a low average annual rainfall (200 mm or less) — in reality, annual figures are meaningless as rainfall in Australian deserts tends to be very sporadic, unpredictable and localised;
- they experience high evaporation rates and large seasonal as well as diurnal temperature variation; and
- they are incapable of sustaining permanent pastoral or agricultural occupation because their type of plant cover is either unpalatable to stock or insufficient to support pastoral or agricultural use.[1]

Why Is the Great Victoria Desert Special?

The Great Victoria Desert has many characteristics which set it apart as a fascinating and exceptional region of Australia:

- It is the largest dune desert in Australia.[4]
- There are no major watercourses which flow through the desert's centre.

Peter Canty

LEFT: Early morning light filters through a black oak Casuarina pauper *woodland in the Yellabinna region of the Great Victoria Desert.*

BELOW: A rainbow, set against a threatening sky, highlights the contrasting colour and mood of the desert landscape.

PREVIOUS PAGE: The wilderness quality of the Great Victoria Desert is unquestioned.

George Wood

- Underlying the dunefield there is an ancient land whose basement rocks have a close affinity with rocks in Antarctica, and whose former surface has been dissected by relic palaeorivers and drainage channels.
- Despite an arid climate, the Great Victoria Desert is remarkably well vegetated and features diverse eucalypt, casuarina and acacia communities. The tree most characteristic of the desert is the spectacular marble gum *Eucalyptus gongylocarpa*.
- The Great Victoria Desert is one of the richest areas for reptile fauna in Australia, while some of Australia's rarest and most beautiful birds such as the scarlet-chested parrot *Neophema splendida* and Alexandra's parrot *Polytelis alexandrae* are found here.
- The Great Victoria Desert acts as a biogeographically active corridor linking mallee-inhabiting fauna from south-eastern and south-western Australia.
- The desert's heartland has never been grazed by stock and is almost entirely free of introduced plants.
- The Great Victoria Desert is one of the most sparsely populated areas in Australia.
- The Great Victoria Desert occupies a substantial portion of the traditional lands of Aboriginal people from the Western Desert Culture Bloc (in particular, the Ngaanyatjarra, Yankunytjatjara, and Pitjantjatjara communities). There is a strong spiritual bond between the desert and its traditional custodians, and the desert is intimately associated with a number of Dreaming tracks and sites of cultural importance.
- Perhaps the greatest attribute of the Great Victoria Desert is its relative ecological integrity — it has an undisputed wilderness character whether measured subjectively by its essential naturalness and atmosphere of vastness and solitude, or objectively by its remoteness from human intervention such as land disturbance or roads.[4]

Chapter 1

Biophysical Description

Climate

Detailed and accurate weather data from the centre of the Great Victoria Desert have never been recorded, rather they have been inferred from records collected at weather stations located around the desert periphery, such as Maralinga, Giles, Ernabella, Ceduna, Coober Pedy and Cook.[1–3] Nevertheless those who visit the desert will testify there is considerable daily and seasonal variation in temperature. Day temperatures are warm to very hot (30–45°C) in summer and mild (15–25°C) in winter. Night temperatures may drop to around or below freezing point in winter. Mean annual evaporation rates are high and vary from 2.5 to 3 metres in the southern part of the desert to near 4 metres on the desert's northern edge.[1] The mean annual rainfall of the Great Victoria Desert is officially classed as low (ranging from 150 mm in the south to 200 mm in the north). However, in a practical sense the desert's rainfall is unpredictable, erratic and highly variable from year to year. Rain in the Great Victoria Desert comes from three main weather patterns. Winter rainfall is associated with frontal activity across southern Australia. Summer rains may occur from either isolated thunderstorms or rain-bearing depressions which represent the tail end of cyclonic or tropical/monsoonal activity in northern Australia. The latter rains tend to be heavy.

When considering the effect of climate on a region, particularly in relation to the vegetation it supports, rainfall and temperature should not be considered as independent variables. The concept of bioclimate — a synthesis of rainfall, its amount and seasonal distribution, and temperature — has been widely promulgated and applied to produce climatic charts (ombrothermic diagrams) for an area/region.[4] Monthly rainfall (in millimetres) and temperature (in degrees Celsius) are both plotted on such charts, with the scale for rainfall double that of temperature (R = 2t). For areas such as the Great Victoria Desert region which experience a 'desert climate', the rainfall line will always fall below the temperature line and all months of the year will be classified as 'dry'. This does not imply rain does not fall in the area, but rather that precipitation is generally inadequate to promote regular plant growth and there is no assured growing season.[4]

Geological History

The Great Victoria Desert is underlain on its eastern, western and northern margins by an ancient crystalline basement comprising rocks at least 1000 million years old (Table 1). On the Western Australian side of the desert, these rocks form the Yilgarn Block of the Western Shield and consist mainly of granites (as well as gneiss and greenstone) aged between 2900 and 2500 million years old.[5] In South Australia, the eastern part of the desert overlays the Gawler Craton (or shield), whose basement rocks vary between 2600 and 1450 million years old.[6] The northern edge of the desert is rimmed by the Musgrave Block, a major tectonic unit of Central Australia. The Musgrave Block is composed of gneiss, granite and gabbro (a coarse-grained rock chemically similar to basalt, and dark in colour) which were probably formed between 1800 and 1000 million years ago.[6]

Repeated faulting and sagging over millions of years have resulted in the formation of depressions or basins, which have been slowly filled with regionally horizontal layers of sediments. These sediments were initially derived from erosion of the ancient crystalline landscape around the basin margins, then washed into the basins by ancient rivers and spread by a succession of transgressing and receding seas. Four such sedimentary basins form an important part of the geological record of the Great Victoria Desert region: the Officer, Arckaringa, Eromanga and Eucla Basins.

The Officer Basin underlies most of the Great Victoria Desert across both Western Australia and South Australia. This large, deep, east-west orientated basin was formed between approximately 800 and 350 million years ago. Its sediment consists predominantly of sandstone, but also of siltstone and shale and, to a lesser extent, limestone and halite (salt).[6]

The Arckaringa Basin, which underlies the desert's eastern edge, was formed as a depression on top of the Officer Basin around 300–250 million years ago. It comprises both glacial and marine sediments (fine sands, mudstone and siltstone) as well as coal.[2]

The desert's rainfall is low and very sporadic. The heaviest falls of rain result from thunderstorms during the summer months.

Peter Canty

Cobble speads of weathered calcrete or silcrete may often be found in dunefield corridors of the Great Victoria Desert.

Mark Shephard

Geological record of the South Australian section of the Great Victoria Desert (adapted with permission from Mines and Energy SA).

TABLE 1: Geological time scale

Era	Period	Epoch	Million years ago
Cainozoic	Quaternary	Holocene (Recent)	
			0.01
		Pleistocene	
			1.8
	Tertiary	Pliocene	7
		Miocene	26
		Oligocene	36
		Eocene	50
		Paleocene	65
Mesozoic	Cretaceous		
			135
	Jurassic		
			195
	Triassic		
			230
Palaeozoic	Permian		
			280
	Carboniferous		
			350
	Devonian		
			400
			540
Precambrian	Proterozoic		
			2500
	Archaean		
			4600

The Eromanga Basin was formed around 200 million years ago. It was filled initially with freshwater deposits and later by marine sediments, when much of the continent was traversed by a great inland sea.

Rifting of Australia and Antarctica prior to around 100 million years ago led to the formation of basins along the Australian continent's newly formed southern margin. Later subsidence resulted in the Eucla Basin being formed on the southern edge of the desert. Non-marine conglomeratic sandstone, siltstone and shale were initially deposited in this basin.[7] As Australia moved further north of Antarctica, major global sea-level fluctuations resulted in regular inundation of the Eucla Basin with marine sediments, particularly limestone. The sea retreated for the last time around 15 million years ago as the northern part of the basin became uplifted to form land. The present-day land surface is now considered a modified sea floor and forms the vast, treeless Nullarbor Plain, one of the oldest and largest karst (eroded limestone) landscapes in the world.

Fossil river systems developed perhaps as early as 100 million years ago and certainly by 45 million years before the present. They flowed from headwaters in the Musgrave Block, Western Shield and Gawler Craton, and etched a winding path across basin surfaces (in a southerly and south-easterly direction) towards the

Nullarbor Plain area. In subsequent geological time, whenever a wetter climatic period occurred (for example around 50–36 million years ago and 1.8 million years ago), these palaeorivers would be reactivated and flow again.[8] Today, one of the most stunningly visual reminders of these ancient river systems is the Serpentine Lakes.

As well as the deposition of sediments in a succession of basins, episodic deep weathering and surface induration during the Cainozoic Era (see Table 1) helped shape the desert's pre-existing land surface.

After rifting and around 65 million years ago, the Australian landscape entered a long period of tectonic stability, which resulted in its surface remaining relatively flat and undisturbed. The combined effect of tectonic stability and periods of high rainfall caused deep chemical weathering which affected both older basement rock and younger sedimentary rocks of the Officer, Arckaringa and Eromanga Basins.

About 35–26 million years ago the Australian climate became drier, and hard cappings, or duricrusts, developed on the exposed land surface. These cappings formed variously of silicon oxide, iron oxide and calcium carbonate (later). A duricrust surface rich in calcium carbonate is termed calcrete. If the duricrust is

TABLE 2: Land system classification within the South Australian section of the Great Victoria Desert[10,11]

Province	Environmental region	Major environmental associations (EA)	Code[a]	EA area (million hectares)	Description of landform
Northern Arid	Western Sandplain	Victoria Desert[b]	8.2.1	4.62	Extensive dunefield with silcrete rises and shallow depressions
		Nurrari	8.2.2	0.66	Plain with dunes and a chain of dry saltlakes
		Giles	8.2.6	1.16	Gently undulating plain with sand dunes and gibber-covered rises
		Dingo	8.2.7	1.29	Undulating plain with dunes, low gibber-covered rises and shallow, sandy depressions
		Purndu	8.2.8	2.06	Gently undulating dune-covered plain
Western Pastoral	Great Victoria Desert	Yellabinna	7.5.2	4.36	Plain with closely spaced dunes

[a]The map on page 22 shows the location of Environmental Associations, and their code numbers.
[b]There are up to 10 other Environmental Associations whose boundaries lie within or near the edge of the Great Victoria Desert. Table 2 lists only some of the more important ones.

Mark Shephard

ABOVE: The vast dunefields of the Great Victoria Desert create a sense of space and solitude for the desert traveller.

Location of selected Environmental Associations in the South Australian section of the Great Victoria Desert (adapted from Laut[10,11]).

Mark Shephard

Simon Coxon

*ABOVE: The desert's sandplains are often vegetated by marble gum woodlands with a spinifex (*Triodia*) understorey.*

RIGHT: The Serpentine Lakes wind their way along the South Australian–Western Australian border and are a present-day visual reminder that palaeorivers once flowed across this ancient land surface.

BELOW: On the southern fringe of the Great Victoria Desert, dunefields merge with the vast Nullarbor Plain (visible as the featureless section of land in the centre left of the photograph).

Peter Canty

composed primarily of silicon oxide then it is called silcrete; while if iron oxide is the major constituent, the duricrust is known as ferricrete.[9] In the Great Victoria Desert today, areas of both calcrete and silcrete are widely exposed as low rises between the dunefields. Extensive areas of ferricrete occur in the north-western part of the desert in Western Australia and near Emu in South Australia.

Major short-term climatic oscillations were subsequently experienced during the Quaternary as polar ice caps grew and then melted. The most recent ice age peaked around 18,000 years ago. In Australia, the ice ages resulted in cold, dry, more intense winds and relative aridity, while interglacial periods had wetter climates, similar to present-day conditions.

The sandplains of the Great Victoria Desert were formed during the glacial periods when strong winds redistributed sands eroded from outcropping crystalline basement and sedimentary rocks. Under the influence of predominantly westerly winds, much of the sand was reworked and built into the longitudinal west-east dunefields of the Great Victoria Desert. The exact age of the dunefield's sands is uncertain and the subject of much current discussion. Elsewhere in the continent, geologists date comparable dunefields from between 300,000 to as little as 10,000 years old.

Environmental Regions

The area occupied by the Great Victoria Desert is shared equally by Western Australia and South Australia. The major environmental regions of both States (including the Great Victoria Desert) have been classified according to two different systems.

Western Australia is divided into an hierarchal structure of Natural Regions comprising Provinces, Regions and Subregions. In Western Australia, the Great Victoria Desert Region covers an area of 21 million hectares.[4] Its boundaries have been derived from vegetation mapping and coincide closely with the former Helms Botanical District. The Great Victoria Desert Region forms part of the large Eremean Province which embraces the State's arid interior and constitutes around 70 per cent of the State by area.

In South Australia, Landsat imagery has been used to divide the State into Environmental Provinces, Environmental Regions and Environmental Associations.[10,11] The area occupied by the Great Victoria Desert and its periphery overlaps two major Environmental Provinces, namely Northern Arid and Western Pastoral (see Table 2, and the map on page 22).

A national integrated approach to land system classification (in which Australia is being divided into a series of Biogeographic Regions that are independent of State boundaries) is currently being devised. The Great Victoria Desert Bioregion forms part of this proposed new classification.

Landforms

Sandplains and Dunefields

Sandplains and dunefields are the dominant landforms of the Great Victoria Desert. In broad cross-section, gently undulating sandplains are common on the western edge of the desert in Western Australia, dunefields are centred on the area either side of the

South Australian–Western Australian border, while sandplains and areas free of dunes become prevalent again further east.

The Great Victoria Desert's dunefield forms the southern part of a vast anticlockwise whorl of dunefields across the Australian interior (which includes those of the Simpson and Great Sandy Deserts).

The dunefields comprise sandridges and the area between the ridges known as the interdune corridor or swale. In general the sandridges are longitudinal, narrow and crested, and run parallel to each other in an overall west-east direction. They range from 5 to 20 metres in height and may run unbroken for up to 100 kilometres.[12] Interdune corridors are generally 300–800 metres wide, with higher dunes having wider corridors.

Sandridge pattern and orientation is not as regular as that in the Simpson Desert and local variation is common, reflecting the influence of underlying topography and the contribution of winds from other directions. Many dunes have twin crests while some converge to form 'V'-junctions which point north-east, east or east-south-east. Clustered, irregular, ill-defined network dunes occur north of Ooldea. Dunes in the far western part of the Yellabinna Environmental Association trend south-west to north-east, although south of the Ooldea Range they are aligned in a west-north-west/east-south-east direction. A northerly wind component may account for the encroachment of dunes over margins of the Nullarbor Plain.[2] The mobile sandridge crests are composed almost entirely of loose sand grains that are pale orange to orange-red in colour due to staining with iron. The older sands of the dune core and swales are red-brown, or brown to orange coloured, and cemented to varying extents with calcrete.

Playa Lakes

Playa lakes are a minor, but locally significant landform in the desert. They occur in topographically low-lying regions and many represent the dried remnants of former drainage channels which flowed in wetter times.

The Serpentine Lakes system comprises a chain of lakes which run in a north-south direction for more than 100 kilometres along the Western Australian–South Australian border. Their sinuous path follows the main channel of a major palaeoriver, and today the lakes constitute a significant landscape feature of the Great Victoria Desert. The Anne Beadell Highway, the main east-west access route across the desert, traverses one of the larger lakes. Low cliffs up to 10 metres high fringe the shoreline and provide excellent views of the lake. The exact age of the Serpentine Lakes' palaeochannel is unclear but certainly the oldest preserved sediments taken from the related drainage are between 36 to 50 million years old. The channel has undoubtedly flowed again since this time, as evidenced by the recent discovery of an outflow of sediment from the channel on the edge of the Nullarbor Plain. The sediment, which has formed an alluvial fan, is younger than the Nullarbor Limestone over which it has been deposited. The Serpentine Lakes also feature heavily in Aboriginal mythology and are known as Mamungarintja by traditional elders of the region (see Chapter 12).

Other playa-lake systems in the desert include the Nurrari and Wyola Lakes and Lakes Dey-Dey, Maurice and Bring in South Australia, and the Forrest, Wanna,

Mike Donaldson

Pulpit Hill is located on the edge of the Morton Craig Range in the western Great Victoria Desert and represents a fine example of a flat-topped hill.

The north-western Great Victoria Desert features spectacular breakaway country.

Peter Canty

AR Martin, Mines and Energy SA

Mark Shephard

TOP: Mt Finke, in the south-eastern Great Victoria Desert, is a breathtaking sight from the air.

ABOVE: The Everard Ranges are located at the eastern extremity of a series of ranges which mark the northern limit of the Great Victoria Desert.

Carlisle, Plumridge and Yeo Lakes in Western Australia. A string of lakes, including the Yarle and Choolalie Lakes, is also found among the network dunes between the Ooldea Range and the Nullarbor Plain.

The exposed surface of these lakes generally consists of dry clay, silt or sand, veneered with a salty or gypsiferous crust.[12] Many playa lakes have lee-side dunes (lunettes) on their eastern margin.

Small claypans and salinas may also form in interdune corridors as a result of ephemeral runoff from dunes.

Other Landforms

Low, stony rises capped by hardened calcrete, silcrete and ferricrete are also present in the desert. They may form extensive undulating plain surfaces in areas devoid of dunes, or may appear as cobble spreads if broken down by weathering and erosion.[12]

Numerous isolated flat-topped hills and extended breakaway ridges are prominent landscape features in the north-western Great Victoria Desert and along the Connie Sue Highway between Neale Junction and Warburton. Breakaways represent erosional fronts between an old plateau surface capped with ferricrete (on high ground) and a young rejuvenated plateau which receives local runoff (below the breakaway). The breakaway edge recedes slowly over millions of years, as softer layers, less resistant to weathering processes than ferricrete, are undercut.

The northern edge of the Nullarbor Plain, one of the world's largest marine karst landscapes, is buried beneath the dunes of the Great Victoria Desert (particularly in the Yellabinna area).[2,8] The plain's surface is pock-marked by shallow circular depressions with gentle slopes and flat clay floors. These so-called 'dongas' are solution features formed by the ponding of run-off from depressions in the limestone surface.[13]

The Ooldea, Barton and Paling Ranges are a narrow chain of rises found near the Maralinga area. The Ooldea Range is connected to the Barton Range by the relatively short Paling Range.[8] They were formed around 35 million years ago as coastal dune systems along the shoreline of a former sea. There are no other sand dune structures of this age anywhere in Australia, making these 'ranges' a unique landscape feature. Once formed they provided an effective barrier to any further northward incursion of the sea. The ranges are composed of quartz sand up to 175 metres thick and have a calcrete capping.

Tietkins Plain, a flat karst feature similar to the Nullarbor Plain, occurs on the northern side of the Ooldea Range. In contrast to the marine limestone substrate of the Nullarbor Plain, Tietkins Plain was formed as a result of the deposition of freshwater limestone and dolomite around 5–15 million years ago.

Mt Finke is an impressive landscape feature in the south-eastern corner of the Yellabinna Region. Its surface is bevelled and has developed on resistant quartzite. The Everard, Musgrave, Mann, Tomkinson, Birksgate, Jameson, Blackstone and Warburton Ranges rim the desert's northern margins.

Chapter 2

A Veritable Botanic Garden

The often misguided concept that Australian deserts are arid wastelands devoid of life is completely dispelled by the outstanding and impressive biological diversity found amongst the plants and animals of the Great Victoria Desert. The region simply teems with life! For an area with such a low and unpredictable rainfall, the desert landscape is remarkably well vegetated. For those with a special interest in plants, the Great Victoria Desert is a veritable botanic garden.

The Desert Vegetation

The Western Australian component of the Great Victoria Desert is known as the Helms Botanical District.[1] Richard Helms was the naturalist on the 1891 Elder Scientific Exploration Expedition which crossed the desert from its north-eastern edge in South Australia to its south-western corner near Queen Victoria Spring in Western Australia.

Queen Victoria Spring, named by Ernest Giles in 1875, and its surrounding Nature Reserve has a remarkable history of botanical collection spanning more than 100 years.[2] The spring itself was an important source of water for numerous exploring, surveying and prospecting parties who ventured into and around the Great Victoria Desert and neighbouring Nullarbor Plain between the mid-1870s and the 1920s (see Chapter 7). As a result, the region's flora was better collected in the early years of settlement than any other desert locality in Western Australia.[2] This historical connection is reflected in the scientific names of many local plants, for example *Calothamnus gilesii, Banksia elderiana* and *Eriachne helmsii*.[2]

Since 1953, Queen Victoria Spring Nature Reserve has been visited by a variety of prominent botanists and naturalists, including John Beard, the Serventy brothers, Harry Butler, members of the Western Australian Herbarium and the CSIRO. David Pearson, from the Department of Conservation and Land Management (CALM), has recently compiled a flora list for the Nature Reserve, prepared a vegetation map of the area showing the distribution of vegetation communities, and examined the fire history of the region.[2]

Queen Victoria Spring Nature Reserve is of considerable botanical importance because it is situated at the interface between the flora of the Great Victoria Desert and the more mesic south-west of Western Australia. Desert acacias, mallees and marble gum (*Eucalyptus gongylocarpa*), characteristic of the sandplains and dunefields of the Great Victoria Desert, coexist with genera such as *Banksia* and *Conospermum*,

David Pearson

The swordfish banksia Banksia elderiana *is the only banksia species whose distribution extends into the Great Victoria Desert in Western Australia. The species name* elderiana *refers to the Elder Scientific Exploration Expedition, which visited Queen Victoria Spring in the south-western Great Victoria Desert during 1891.*

whose affinities lay in south-western (and eastern) Australia. The reserve also contains one species of declared rare flora, *Conospermum toddii*, and several which are locally endemic.[2]

The main plant communities occurring in Nature Reserves located in the central and western parts of the Great Victoria Desert in Western Australia have also been documented (Table 3).[3,4] A vegetation map for the entire Western Australian section of the desert was produced in 1974.[5]

In South Australia considerable botanical work has been conducted in the desert over the past 25 years. In the winter of 1967, the South Australian Pastoral Board carried out an inspection of more than 10 million hectares of land in the far west and far north-west of the State in an attempt to assess its pastoral potential.[6] Their

TABLE 3: Major vegetation communities in the Neale Junction and Yeo Lake Nature Reserves[3,4]

Reserve	Location in desert	Total area (hectares)	Description of community	Percentage of total Reserve area occupied by the community
Neale Junction	Central	725,000	*Eucalyptus gongylocarpa* low open woodland between sandhills	46
			Eucalyptus gongylocarpa low open woodland	31
			Acacia aneura low woodland between sandhills	10
			Acacia aneura low woodland	7
			Mallee tall shrubland	6
			Saltbush and samphire low open shrubland	1
Yeo Lake	Western	320,000	*Eucalyptus gongylocarpa* low open woodland over *Eucalyptus youngiana* tall shrubland over *Triodia basedowii* open hummock grassland	51
			Acacia aneura low woodland	30
			(Salt lakes)	12
			Saltbush and samphire low open shrubland	4
			Mallee tall shrubland	3

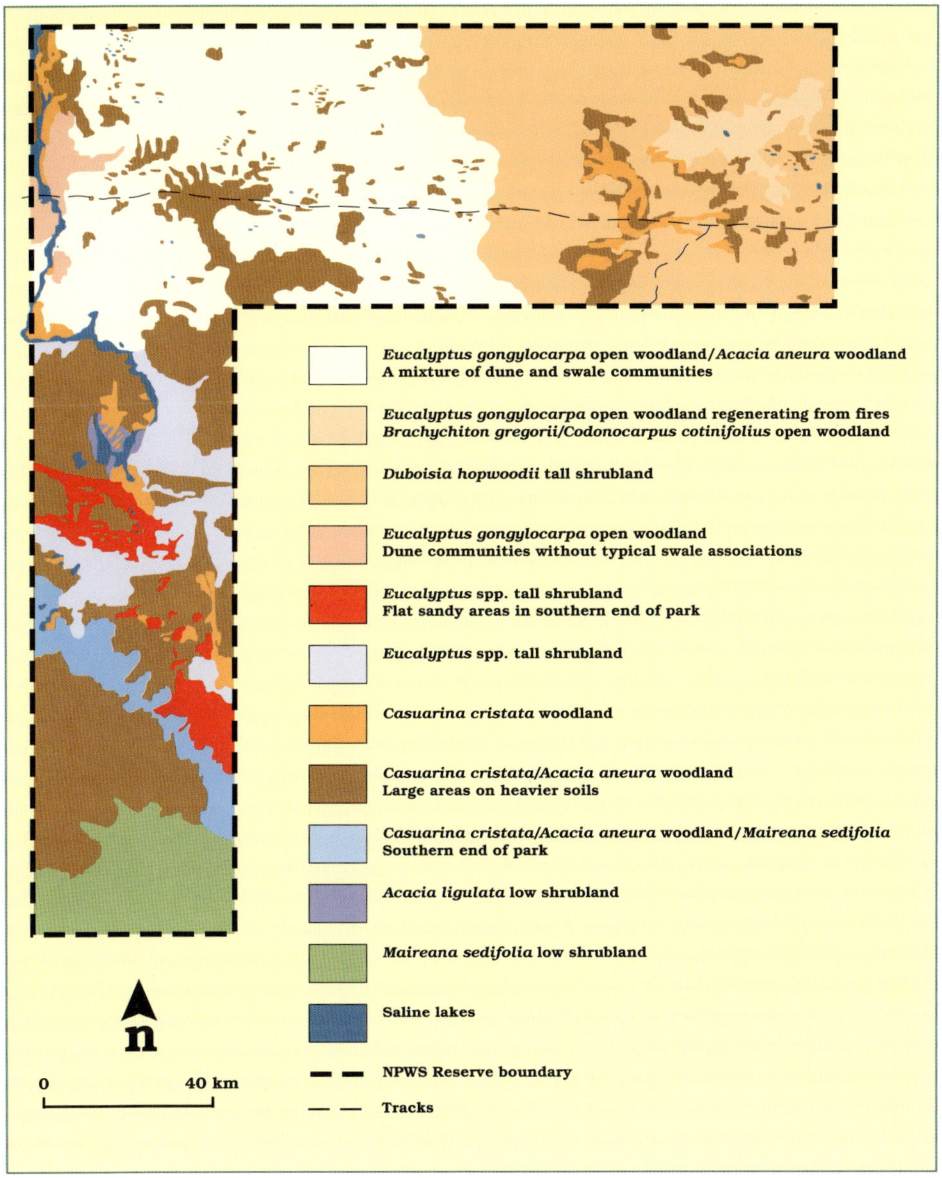

A vegetation map featuring the major plant communities in the Unnamed Conservation Park, South Australia, adjacent to the Western Australian border. (Source of data: Ecological Survey Unit, Department of Environment and Planning, 1979. Map digitisation and production: Information Systems Branch, Department of Environment and Planning, 1992.)

broad reconnaissance covered almost the entire area occupied by the Great Victoria Desert in South Australia, and included valuable descriptive records of vegetation along two seismic traverses (the Spanner Highway and Serpentine V2 line) which are now overgrown and no longer accessible by vehicle. The Spanner Highway linked Emu to Officer Creek in the north-eastern part of the desert, while the Serpentine V2 line ran north from the Anne Beadell Highway and joined the Mt Davies Road in the north-western corner of the State. The survey also included the bulk of the area now known as the Maralinga Tjarutja Lands which lie between the Unnamed and Tallaringa Conservation Parks (see Chapters 11 and 12). The vegetation of this area has not previously or since been comprehensively researched. A vegetation map, comprising six basic plant associations, was prepared from the Board's initial field inspection and refined three years later following a return visit to selected areas for more detailed examination.

Since this broad-scale survey, numerous other studies have been conducted on more specific, localised regions of the desert.

In 1979, the Ecological Survey Unit of the former Department of Environment and Planning undertook a vegetation analysis of the Unnamed Conservation Park adjacent to the South Australian–Western Australian border.[7] Landsat imagery and ground surveys were used to produce a vegetation map consisting of nine major plant communities.[8] Within the park, 274 species of flowering plants were collected, including four officially classed as rare or threatened (*Calandrinia disperma, Darwinia micropetala, Frankenia cineria and Frankenia mucosa*).[9]

The Nature Conservation Society of South Australia undertook a major study of the flora and fauna of the Unnamed Conservation Park in 1980. Influence analysis was used to define groups of plant species that were associated. These associations were then related to particular landforms. The presence of the dune–interdune sequence and the location and depth of calcrete and carbonate hardpans were two important features influencing plant distribution.[10]

In 1992, an assessment of the occurrence of fire and its effect on plant communities of the Unnamed Conservation Park was conducted by Josephine Morelli, on behalf of both the South Australian National Parks and Wildlife Service and the Australian National Parks and Wildlife Service (now the Australian Nature Conservation Agency).[11] As part of this project, 61 plant species representing 25 families were collected primarily at 10-kilometre intervals along the Anne Beadell Highway across the northern part of the park. Seven major vegetation communities (similar to those described by the Ecological Survey Unit, 1979) were identified within this section of the park. These were marble gum open woodland, marble gum open woodland/mulga woodland, mulga woodland, mulga woodland/black oak woodland, mallee tall shrubland, tall shrubland, and low saline shrubland.

A major study of the vegetation of the eastern Great Victoria Desert was carried out in 1988 on 1.2 million hectares of land now reserved as Tallaringa Conservation Park. Aerial reconnaissance was used to photograph broad-scale variation in vegetation, while a ground survey involving systematic vegetation sampling was conducted at 97 sites. Two hundred and five perennial plant species were recorded, incorporating nine major vegetation types spread across a 150-kilometre north-south environmental gradient.[12] None of the nine vegetation alliances had been previously conserved within the South Australian national park system. The Tallaringa Park contains regenerating mulga woodland and serves as a benchmark against which the impact of active pastoralism in mulga lands located further east and south can be compared. A floristic vegetation map of the Tallaringa area was also produced as a result of the study.[12]

A follow-up survey was conducted in October 1993 by a party from the Melbourne-based Australian and New Zealand Scientific Exploration Society

(ANZSES). Other minor surveys were carried out in the Lake Phillipson and Wilkinson Lakes area to the south and west of Tallaringa during the 1980s.[13,14]

The Yellabinna Environmental Association in the south-eastern section of the Great Victoria Desert was the subject of a detailed biological survey during 1987.[15] There are 686 plant species now known from the Yellabinna area, 187 (27 per cent) of which were first collected in 1987. Twenty-three vegetation communities occurring on six major substrate types were mapped during the survey. Yellabinna was shown to be a significant area for the conservation of plant biodiversity. It is a major link or biological corridor connecting areas of mallee and acacia scrublands/woodlands in the Eyre Peninsula region of South Australia with those in the south-west of Western Australia.

An overall comparison of the flora of the Great Victoria Desert with that of the better known Simpson Desert provides some interesting similarities and notable differences. The three most successful plant families in both deserts are the same: Compositae (composites or daisies), Chenopodiaceae (saltbushes, bluebushes) and Graminae (grasses). The family Myrtaceae (notably eucalypts) is widely represented in the Great Victoria Desert, but poorly in the Simpson Desert. The reverse is true of the family Leguminosae (pea-like flowers). The genus *Acacia* (wattles) has a similar proportional representation of species within both deserts, but the number of species in common is few. Lobed spinifex *Triodia basedowii* and horse mulga *Acacia ramulosa* occur widely on the dune crests of the Great Victoria Desert sandridges; in contrast, spinifex is restricted mainly to the interdune corridors of the Simpson Desert's dunefields. Sandhill canegrass *Zygochloa paradoxa*, the most abundant plant on the dune crests of the Simpson Desert, is thought to be absent from the Great Victoria Desert.

Plants of Special Interest

The most distinctive tree in the Great Victoria Desert is the marble gum *Eucalyptus gongylocarpa*. It is also commonly known as the desert gum, bara gum or baarla. Marble gum grows on areas where deep sandy soil occurs, notably sandplains on the Western Australian side of the desert, and sand dunes at the northern end of the Unnamed Conservation Park extending north and east across the Anangu Pitjantjatjara Lands.[11,16] Visitors to the desert who travel along the Anne Beadell Highway between Vokes Hill Corner and Neale Junction will regularly encounter corridors of this impressive and handsome tree, whose presence gives the desert a unique character. On a recent private trip to the desert, I recorded the marble gum in South Australia as far north as 27°32'29"S/129°51'44"E. It is not found in the Yellabinna or Tallaringa regions of the desert.

Sandplains may often comprise a low diversity of plants other than marble gum and a lobed spinifex understorey. In sandhill country, marble gum communities are generally mixed with mallee and a mosaic of other trees and shrubs, including mulga woodlands in wider interdune corridors which have a more clay base.[5]

The marble gum usually grows as a single-stemmed tree up to 20 metres high, but some may have a crooked and twisted appearance. Its bark is smooth, white and mottled with small peeling red-brown flakes.[17] The leaves are short, narrow, arranged in opposite pairs and may have a slight waxy texture. Club-like buds occur in spreading clusters, while its small almost spherical-shaped fruits give rise to delightful pale white flowers.[18] Marble gum was first collected by Richard Helms in early July 1891 at the Elder Expedition's Camp 17, 40 kilometres south-south-west of Mount Watson in the Birksgate Range on the desert's northern fringe.

Mallee is a collective term for those eucalypts which have several (multiple) stems arising from a common rootstock (or lignotuber) running horizontally below the ground surface. Mallee is thus a particular growth form of eucalypts. In the Great Victoria Desert, mallees are widespread and may occur in association with marble gum open woodlands, or as belts of tall shrubland which often comprise small associations of different species of mallee.[8]

One of the desert's most common and attractive mallees is the Ooldea mallee

Mark Shephard

ABOVE AND BELOW: The marble gum Eucalyptus gongylocarpa *displays many growth forms. Some trees are relatively straight and single stemmed, while others are twisted and gnarled.*

Chris Wenham

Mark Shephard

ABOVE: The large-fruited Ooldea mallee Eucalyptus youngiana *is a conspicuous element of mallee shrublands in the Great Victoria Desert. It displays spectacular red, cream or golden yellow flowers.*

RIGHT: A small grove of the Christmas tree mulga Acacia aneura *var.* conifera *may be found 1 km west of Vokes Hill Corner in the Unnamed Conservation Park.*

Mark Shephard

Eucalyptus youngiana. It is found across much of the desert's sandplain and dunefield country. In the northern part of the Unnamed Conservation Park and along the Anne Beadell Highway it is often co-dominant with marble gum open woodland on sandy dune slopes and swales. The Ooldea mallee is straggly in appearance and has rough, loose, dark brown bark along much of its stem. Its buds are very large, being wide and short in length. Both the torsus and cap are ribbed. It has strikingly large, bright, showy flowers which may be cream, golden yellow or red in colour and may appear at any time between July and November after sufficient rain.[18] Characteristically, the Ooldea mallee's woody mature fruit is considerably larger than those of other mallee species.[7]

The kingsmill mallee *Eucalyptus kingsmillii* subsp. *alatissima* has large colourful flowers and a substantial woody fruit similar to that of the Ooldea mallee.[19] It is relatively common in an area located half way between Neale Junction and the Serpentine Lakes.[19] In South Australia it is known only from the northern extremity of the desert (near the Birksgate Range) and from the Serpentine Lakes region towards Vokes Hill Corner.[18] The kingsmill mallee may be distinguished from the Ooldea mallee by its buds which have prominent winged ribs, and the size of its mature fruits which are slightly smaller than those of the Ooldea mallee.

Other interesting mallees found along the Anne Beadell Highway between the eastern boundary of the Unnamed Conservation Park and the Serpentine Lakes include Ewarts mallee *Eucalyptus ewartianna*, summer red mallee *E. socialis* and kopi mallee *E. striaticalyx*. The Vokes Hill mallee *E. eremicola* and the Victoria Desert mallee *E. concinna* may also be found in this general region and extend well into Western Australia.[17] Further to the south, tall shrublands of red mallee *E. oleosa* and the pimpin mallee *E. pimpiniana* occur. In the Yellabinna region a variety of other mallees including the slender-leafed red mallee *E. leptophylla*, Yumbarra mallee *E. yumbarrana*, yorrell *E. gracilis* and the rare Victoria Spring mallee *E. trivialis* have been recorded (as well as many of the aforementioned species). A relatively recent addition to the flora of the Great Victoria Desert is *E. wyolensis*. This mallee or small tree is restricted to patches of red-brown sands over limestone between the northern edge of the Nullarbor Plain and the sand dunes of the Great Victoria Desert.[20]

Numerous species of acacia (wattle) are found in the desert. Mulga *Acacia aneura* var. *aneura* is present as distinct woodland communities in broad interdune corridors where the soil has a greater clay content. It may also occur in combination with black oak *Casuarina pauper* where the soil is clay-loam, or western myall *Acacia papyrocarpa* where limestone-rich soils exist. The ground cover associated with these mulga communities may vary considerably. Pure mulga communities generally have a covering of arid grasses such as *Eragrostis, Aristida* and *Enneapogon*, with ephemeral herbs appearing after rain. Mulga merging with black oak or western myall communities may comprise arid grasses and bluebush, or bluebush and saltbush respectively. Mulga is also prevalent in the north-western and eastern parts of the desert where breakaways and flat open plains covered with a surface layer of ironstone gravel occur. It does not grow on sandy soils of the desert's sandplains and sandridges.

Mulga typically has upright branches which 'point to the sky', needle-like phyllodes or 'leaves', and yellow-spiked flowers which are not often seen. Throughout Australia, nevertheless, mulga occurs in many different growth forms and presents somewhat of a taxonomic nightmare. *Acacia aneura* var. *conifera* is a very interesting form of mulga which is found in the Great Victoria Desert area and has only recently been described. The latter scientific name refers to its unusual conifer or Christmas tree-like appearance.[21]

Other acacia species may dominate tall shrubland communities (not containing marble gum, mallee or mulga). Horse mulga *Acacia ramulosa* is one of the most abundant shrubs found on the desert's dune crests and sandy swales. It is a spreading shrub with many branches rising from slender stems at ground level. The leaves of horse mulga are long and rounded, with numerous fine, parallel, longitudinal veins. Its yellow flower heads are arranged in dense spikes.

Umbrella bush or marpoo, *Acacia ligulata*, is also a common tall shrub of sandy areas. It may be distinguished from horse mulga by its flat, deep green leaves with a central parallel vein and ball-shaped (globular) flowers. A pure community of umbrella bush may be found on the low dunes on the eastern edge of the Serpentine Lakes.

Another distinct vegetation association of the desert is the open woodlands of western myall *Acacia papyrocarpa* with a ground cover of low chenopod shrubs, notably blueblush *Maireana sedifolia* and saltbush *Atriplex vesicaria*. This association, which in places may be mixed with mulga or black oak, occurs mainly on the heavier soils around the north-western edge of the Nullarbor Plain and the south-western edge of the desert. It may also be found to the south-east of Barton and near Emu on the eastern (South Australian) side of the desert. Western myall is a small tree, approximately 8 metres high, typically multistemmed and having a characteristically dark trunk and rounded, umbrella-like canopy.

Other wattles scattered across the desert include dead finish *Acacia tetragonophylla*, which is a gnarled, straggly shrub or small tree whose short needle-like leaves are arranged in sharp, spiky clusters; Giles wattle *A. gilesiana*, a dense, spreading shrub with long, rounded leaves and globular flowers; sandhill wattle *A. burkittii*, a tall shrub with smooth silver-grey branches and long spikes of bright yellow flowers; witchetty bush *A. kempeana*, a dull grey-green shrub with flat leaves containing numerous parallel, longitudinal veins, broad seed pods and yellow flower heads arranged in spikes; and umbrella wattle *A. oswaldii*, a widespread small tree with a single straight trunk, a bushy, rounded canopy comprising long flat leaves, simple globular flower heads on very short flower stalks, and distinctive curling woody legumes.[21]

Apart from eucalypts and acacias, there are many other interesting trees in the desert.

The black oak or belah, *Casuarina pauper*, occurs as woodland communities in areas whose surface is scattered with silcrete or calcrete gibber spreads and underlain by heavier soils. It may also be found on gypsiferous dunes near the Serpentine and Forrest Lakes, in combination with mulga woodland on wider interdune corridors with clay-loam soils, and as open woodland with Victoria Desert mallee at the northern end of the Serpentine Lakes.[7,8]

Black oak is a stately tree with a dark-coloured, rough-barked trunk. Its canopy comprises large branches which in turn support numerous faintly ribbed interlocking branchlets rather than leaves. Its leaves are in fact reduced to a ring of minute scales or teeth which occur at the joints of the segmented branchlets. These scale leaves have evolved as an adaptation to prevent water loss. Like all casuarinas, male and female flowers are confined to separate trees. A compact wooden cone containing winged seeds is found only on female trees. The roots of the black oak, as well as several species of mallee, were used as a source of water in times of severe drought by Aboriginal people. The black oak is displaced on the northern edge of the desert near the Mann Ranges by the better known desert oak *Allocasuarina decaisneana*.

The native pine *Callitris verrucosa* occurs sporadically across the desert, often growing among tall shrubland and woodland communities and favouring deep soils, particularly sand dunes.[1,8] A pure community of this pine was found on a helicopter traverse in the northern sector of the Unnamed Conservation Park.[8] The colour of the leaves can vary from grey-green to dark green.

Bullock-bush *Alectryon oleifolium* is a small tree, 3–6 metres in height, with a broad greyish-green canopy. It grows on sandy soils and is found scattered within black oak and mulga woodlands, and mallee shrublands. This plant is toxic when in flower.[22]

Open woodlands of the beautiful desert kurrajong *Brachychiton gregorii* grow in sand-dune areas. Patches of the tall, erect desert poplar *Codonocarpus cotinifolius* also dominate sand dunes, mainly where marble gum communities have been recently subjected to fire.[8]

The desert kurrajong has a single straight trunk and usually grows to a height of around 8 metres. It has a compact shady crown comprising leaves that are smooth, shiny, hairless and divided into graceful finger-like lobes. Each lobe is long, flat and has a central vein. The kurrajong may shed its leaves during drought.[18] Its

Mark Shephard

Mark Shephard

ABOVE: The smooth, shiny leaves of the desert kurrajong Brachychiton gregorii *usually have three to five finger-like lobes. The kurrajong had many uses for the desert's traditional Aboriginal people.*

RIGHT: The desert poplar Codonocarpus continifolius *is an attractive tree which rapidly recolonises marble gum and mallee communities after fire.*

cream-coloured flowers are bell-shaped and hang in clusters.[23] It has a small beaked woody seed capsule which splits along the centre to reveal around 12 seeds. Kurrajong seeds were traditionally roasted (the taste resembling that of peanuts) and were ground into cakes by Aboriginal people. The seeds were also pounded and mixed with water to make a type of coffee drink by early explorers.[23] The roots of the kurrajong are porous and contain large supplies of water. The kurrajong was therefore a principal water-bearing tree for the Western Desert Aboriginal people during periods of drought. The roots were also chewed to obtain sap and other nutrients.

The short-lived desert poplar is widespread across inland Australia. It resembles the European poplar in general appearance, hence its common name. The botanical name *Codonocarpus* refers to its characteristic segmented, bell-shaped fruit. Like the black oak, its male and female flowers appear on separate trees.

The quandong *Santalum acuminatum* and native apricot *Pittosporum phylliraeoides* are two 'fruits of the desert'. The quandong is a member of the semiparasitic sandalwood family. To propagate, its root system must connect with and draw nutrients from the roots of other plant species. The drooping leaves of the quandong are grey-green, occur in opposite pairs and taper to a short curved point. Its rounded fruit comprises a fleshy, edible red outer layer which forms the main ingredient for quandong jam. The fruit encapsulates a characteristic deeply pitted stone, which is often found at the base of the tree. The stone, in turn, harbours an edible kernel. Emus feed on

Mark Shephard

Mark Shephard

ABOVE: The ripe fruit of the quandong Santalum acuminatum *is 2–3 cm in diameter and consists of a hard, deeply pitted, round stone covered by an edible, red fleshy layer.*

LEFT: The weeping pittosporum Pittosporum phylliraeoides *is widespread throughout inland Australia. Its sticky seeds are encased in a bright, apricot-orange capsule and its pendulous, willow-like leaves are palatable to grazing stock.*

BELOW: Camel poison bush Gyrostemon ramulosus *is a short-lived tree which has a very branched appearance and grows well on the desert's sandy soils.*

Mark Shephard

quandong fruits and help distribute the seeds to other areas by transporting them in their droppings. Another sandalwood found in the Great Victoria Desert is the native plum bush *Santalum lanceolatum*, which has smaller, purple fruits when ripe.

The native apricot or weeping pittosporum is a small graceful tree with the general appearance of a weeping willow. It has long narrow leaves, which are highly palatable to grazing stock, and a hard, round, apricot-coloured seed capsule. The capsule's bitter-tasting, inedible seeds are embodied in a sticky, red pulp which greatly facilitates the dispersal of the seeds by birds. Native apricot is one of the commonest plants which grow in the dongas on the Nullarbor Plain to the south of the desert.

Camel poison bush or wheel fruit, *Gyrostemon ramulosus*, is so-named because its leaves are purported to be toxic to camels. The nature of the leaf toxin is currently being investigated by Dr Brian Gatehouse at Monash University. Camel poison bush is a small tree which has a grey corky bark and many tightly clustered, small branches. It has long, slender, dark-green leaves and a characteristic pear-shaped fruit divided into many segments. Male and female flowers occur on separate bushes.

Grevilleas and hakeas are not well represented in the desert. Perhaps the most widespread is the desert grevillea *Grevillea juncifolia*, with its long rush-like leaves, spectacular golden-orange flowers and papery seed pods. The sandhill spider flower or rattlepod grevillea, *G. stenobotrya*, also has long narrow leaves, but is distinguished from the former species by its creamy-white flowers and its seed case which 'rattles' in the wind. Similar to scooping honey out of a jar, the flowers of both species of Grevillea were most likely chewed by Aboriginal people to provide a source of nectar and sugar. *G. nematophylla* is a tall shrub which grows on sandplains and is often associated with mulga. The lateral roots of this shrub were a source of edible grubs for the desert's Aboriginals. The desert's hakeas include needlewood *Hakea leucoptera*, corkwood *H. suberea* and *H. francisiana*; they may be differentiated from grevilleas by their hard woody seed capsule.

Some of the desert's more interesting shrubs include the blackboy or desert grass tree *Xanthorrhoea thorntonii*, pituri *Duboisia hopwoodii* and desert thryptomene *Thryptomene maisonneuvii*. Blackboys, with their solid, black trunks and long, stiff, pointed leaves occur as scattered colonies on sandplains in the south-western part of the desert, notably near Queen Victoria Spring. They are also found among desert oaks south of the Mann Ranges on the northern extremity of the desert.[1,16] Pituri rapidly recolonises and stabilises areas of dunefields where marble gum and mallee communities have recently been disturbed by fire.[8] Aboriginal people widely use this plant as a narcotic by combining the dry crushed leaves of pituri with the ash of other plants, such as umbrella bush, and then chewing the resultant mixture. The delightful sweet-smelling desert thryptomene is a small ericoid shrub with pink or white flowers which may form a dense field or carpet across sandy dune slopes. In the process, thryptomene may successfully out-compete the usual spinifex ground cover or at least reduce it to isolated clumps. The leaves of the desert thryptomene are small, swollen and lack a leaf stalk. Aboriginal people collected thryptomene nectar after frosts. They would shake the shrub's branches which shed the frozen nectar-dew mixture into their wooden bowls (pitis).

A very unusual low shrub found in the desert is the 'upside-down plant' *Leptosema chambersii*. This leafless plant has a mass of prickly, green stems and grows typically between clumps of spinifex. Remarkably, its bright red, showy flowers protrude at ground level from the base of the plant.

Hummock grass or 'spinifex' is the most widespread and dominant ground cover throughout the Great Victoria Desert. Across Australia, the term 'spinifex' is loosely applied to a group of around 30 different species of hummock grasses belonging to two closely related genera, *Triodia* and *Plectrachne*. These grasses, which grow as dense, spiky, rounded clumps or hummocks, cover more than one-quarter of the Australian continent. As a spinifex hummock ages and expands in size, it begins to die in the centre and forms a distinctive growth ring. The rings occur most commonly in old stands and can range from 1 to 6 metres in diameter with only the outer part consisting of living plant material.[22] Lobed spinifex *Triodia basedowii* and porcupine grass *Triodia irritans* are two hummock grasses found commonly in the Great Victoria Desert. Lobed spinifex grows typically as ground cover in marble gum communities. Its leaf blades are hard and leathery (sclerophyllous) and often tightly curled to form sharp, pointed rods. Porcupine grass occurs more often on shallow sandy soils overlying silcrete or clay.[8] *Triodia scariosa* may replace lobed spinifex where *Eucalyptus oleosa* or *Eucalyptus socialis* become the dominant mallee.[1,15] Plectrachne grasses occur on the Western Australian side of the desert while *Triodia lanata* is associated with *Eucalyptus leptophylla* in open mallee scrub in the Yellabinna area.[15] Spinifex grasses are absent from the Tallaringa region. True spinifex (that is, plants belonging to the genus *Spinifex*) occur only on coastal dunes, far removed from the Australian desert areas.

A variety of succulent, salt-tolerant, low shrubs may be found along the edge of the desert's saltlakes or salinas. These include the leafless samphires (genus *Halosarcia*), narrow-leafed bluebushes (*Maireana*) and broader-leafed saltbushes (*Atriplex*). These three groups belong to the large plant family known as Chenopodiaceae. This particular type of community is often referred to as a chenopod shrubland.

Natural events such as fire and rainfall may have a dramatic effect on the structure and composition of the desert's plant communities.

Adaptations to Fire

The fire recovery strategies or regenerative mechanisms employed by different plant species and communities in response to a wildfire are varied and interesting. Basically, plants return after fire by either seed or vegetative regrowth (aerial and basal resprouting), or a combination of both.

In marble gum and mallee communities, the build-up of leaf litter around the base of trees and the presence of a flammable spinifex understorey contribute to a high fuel load and may initiate fire. Marble gum will readily resprout after fire from protected, undamaged epicormic buds along its branches and trunk.[11] Mallee regenerates following fire by vegetative resprouting from lignotubers. Fire will also accelerate the release of mallee seeds stored in the canopy and facilitate the establishment of new seedlings.[11] Other plants such as the desert poplar and pituri may opportunistically flourish in recently burnt areas of marble gum and mallee communities.

Ruth Aspley

ABOVE: Hakea francisiana *is a tall shrub with long, narrow leaves and delightful pink-red flowers.*

BELOW: Desert thryptomene Thryptomene maisonneuvii *is an extremely attractive small shrub which flourishes in the desert's sandy terrain. Its small white or pink flowers have five small petals with a stamen between each petal.*

Mark Shephard

David Pearson

The blackboy Xanthorrhoea thorntonii *is common in the Queen Victoria Spring area. Aboriginal people used resin from its trunk as an adhesive and obtained honey from its flowers.*

Mulga is a fire-sensitive species and mature trees are often destroyed by fire. A new mulga community will regenerate in its place through the release of copious, hard-coated seeds which lie dormant in the ground and crack open only after 'hot' fires and heavy rainfalls. A low intensity fire will kill adult trees and is unable to bring about successful seed scarification and release. Mulga also requires a minimum of seven to eight years to attain reproductive maturity. A return fire earlier than this period will cause long-term damage to the community.[11] In general, mulga

Peter Canty

Mark Shephard

ABOVE: Hummock grasses belonging to the genera Triodia *and* Plectrachne *are often found in association with the desert's marble gum and mallee communities, where they dominate the understorey vegetation.*

ABOVE RIGHT: Sturt's desert pea Swainsona formosus *is one of Australia's most spectacular and well-known plants. Fields of flowering desert pea are recorded regularly throughout sections of the Great Victoria Desert after rain.*

RIGHT: The desert traveller who is fortunate to visit the Great Victoria Desert two to three months after rain will be greeted by the sight of brightly coloured wildflowers carpeting the desert landscape.

George Wood

communities are less susceptible to fire than eucalypts, except after heavy rains when a dense ephemeral understorey develops and subsequently dries off. Under their patch-burning regime, Aboriginal people exclude fire from mulga communities to protect important food sources such as grubs, galls, grass seeds and mistletoe.

Black oak woodlands in the southern part of the desert have a relatively low susceptibility to fire because they support a chenopod rather than spinifex understorey. However, communities may be at more risk in the northern section of the desert where a build-up of grasses following rain can occur.[11] The hard rough outer bark of the black oak may provide some protection to inner heartwood from the ravages of fire.

Spinifex constitutes a significant fuel load in marble gum woodlands and mallee shrublands, through which it is the dominant ground cover. Its high resin content and well-aerated, fine leaves are responsible for its high flammability. As spinifex ages and becomes senescent, it is less efficient in carrying a fire. Juvenile triodia hummocks that are less than seven to ten years old are generally incapable of sustaining a fire for long periods due to insufficient flammable fuel.[11]

Response to Rain

Desert rainfall is an irregular and highly unpredictable event. When effective falls do occur, the rain not only stimulates regrowth after fire, but also encourages the dormant seeds of short-lived annual or ephemeral wildflowers to burst into life and transiently carpet the desert floor. This localised 'greening' brings a new freshness, vitality and colour to the desert, and results in a flurry of activity among many of the desert's animal opportunists. The daisy (Compositae) family, in particular, adorns the desert with colourful cascades of yellow, white, pink, lilac, purple and blue flowers. Typical daisies, in which the flower head is surrounded by a ring of ray petals, include the minnie and brachycome daisies (genera *Minuria* and *Brachycome*), burr-daisies (genus *Calotis*) and yellow tops (genus *Senecio*). Paper daisies, whose outer ring of petals is replaced by a row of papery bracts or modified leaves, comprise members of the *Rhodanthe, Chrysocephalum* and *Polycalymma* genera. Billy buttons (*Calocephalus* and *Craspedia* genera) are the best known examples of button daisies, in which the flower head lacks both petals and radiating bracts.

Other desert annuals include parakeelyas (genus *Calandrinia*) and Sturt's desert pea *Swainsona formosus*. Parakeelyas are small succulent plants which have fleshy stems and leaves, and five-petalled purple, pink or white flowers. The Sturt's desert pea is the floral emblem of South Australia. Its brilliant red flowers which feature a shiny black centre are arranged in clusters of three or more and are held erect by strong stems.

Adaptations to Aridity

The desert's plants tolerate aridity by a variety of different adaptations designed to reduce water loss from their leaves. Eucalypts, for example, have flat leaves and present only their edges to the sun. Acacias have evolved flattened or enlarged leaf stalks (phyllodes) which closely resemble and function as leaves. Their stomates (pores through which respiration occurs) are both sunken and reduced in number. Mulga has branches that slope towards the trunk and channel water towards its roots. The black oak has leaves that are reduced to scales and occur only at joints in the drooping branchlets. Sennas (formally known as Cassias) also have modified leaves which are divided into segments or leaflets, while some species of hakea have leaves that are needle-like and terminate in a point. Samphire has fleshy, succulent stems that store water. Some saltbushes are able to excrete excess salt from their leaves. Other drought-resistant plants have leaves that are coated with a 'waxy' varnish (hopbush) or are covered with downy hairs (bluebush).

Chapter 3

Birds

The Desert Birdlife

The first records of the birdlife of the Great Victoria Desert were made by early explorers near the turn of the last century. Ernest Giles and Richard Maurice, who crossed the desert in 1875 and from 1897 to 1903 respectively, made fleeting diary references to the desert's birds.[1,2] Richard Helms, a naturalist with the 1891 Elder Scientific Exploration Expedition, collected 18 birds, 8 nests and 16 eggs from the South Australian side of the desert.[3] Around the time of construction of the Transcontinental Railway Line (1917), birds from the Ooldea region and the Nullarbor Plain were studied in some detail by both prominent ornithologists such as Captain Samuel White, F. Whitlock and A.E. Le Souef, and amateur naturalists including Anthony Bolam.[4–9]

The advent of roads in the mid-1950s to early 1960s allowed better access into and across the desert. Interest in the avifauna of the Great Victoria Desert intensified as a result, and the importance of the region in the evolution and biogeographical distribution of Australia's birds began to emerge. Initially during this period, most of the ornithological field work focussed on the Western Australian sector of the desert. [10–14] Well-known ornithologist Peter Slater recorded birds from the Queen Victoria Spring area in 1955.[10] Naturalist Vincent Serventy also visited the same area in 1956 and 1960.[11] A detailed study of birds was carried out along the Connie Sue Highway in 1965.[12] This road runs due north from Rawlinna on the edge of the Nullarbor Plain through the Great Victoria Desert to Warburton on the southern edge of the Gibson Desert. The road bisects east-west belts of vegetation including open chenopod plains, myall woodland, mallee, marble gum woodlands, mulga, and breakaway country. Seventy-two species were recorded along this transect.[12] Around 10 years later, bird lists were compiled from the Neale Junction, Plumridge Lakes and Yeo Lake areas as part of an assessment of these locations for their suitability as Nature Reserves.[15,16] Birds of the southern desert and neighbouring Nullarbor Plain have continued to be monitored by wildlife managers and scientists, as well as by staff at the Royal Australasian Ornithologists Union's (RAOU) Eyre Bird Observatory near Cocklebiddy.[17,18]

In South Australia most work has concentrated on the Unnamed Conservation Park and Yellabinna region. Extensive surveys were conducted in the Unnamed Conservation Park by members of the Nature Conservation Society of South Australia (NCSSA) in 1980 and the South Australian Ornithological Association (SAOA) in 1990.[19,20] The former project provided an overview of the birdlife of the park and surrounding areas, and examined the habitat preferences of selected birds. Eighty-four species were recorded during this field study (70 from within the park and a further 14 outside its boundaries). A list of 123 species from the eastern Great

Victoria Desert as a whole was compiled from a review of ornithological literature and the records of other observers.[19]

The SAOA study highlighted seasonal changes in both bird abundance and the use of habitat by birds from the region. Sightings of the slaty-backed thornbill *Acanthiza robustirostris*, ground cuckoo-shrike *Coracina maxima* and whistling kite *Haliastur sphenurus* represented first records for the eastern Great Victoria Desert. In addition, the banded lapwing *Vanellus tricolor*, barn owl *Tyto alba*, collared sparrowhawk *Accipiter cirrhocephalus*, redthroat *Sericornis brunneus* and pied honeyeater *Certhionyx variegatus* were not recorded during the NCSSA field survey.[20]

The ground cuckoo-shrike Coracina maxima *was first recorded on the South Australian side of the desert as recently as 1990. It has a distinctive, metallic call and generally lives on the ground in small family parties. Like all cuckoo-shrikes, it gives a characteristic flick of its wings on landing.*

In 1984 another party from the SAOA conducted a bird survey of the Anangu Pitjantjatjara (AP) Lands in the far north-west of the State. Their work centred mainly on the range country (Mann, Tomkinson, Musgrave, Everard and Birksgate Ranges) but also included the northern limit of the Great Victoria Desert.[21] Further studies in this region were carried out by myself in 1993 and 1994, in a project which also incorporated the gathering of traditional ecological knowledge of birds.[22,23]

During 1987, a survey of birds from the Yellabinna region in the desert's south-eastern corner recorded 88 species.[24] The 12 most commonly recorded birds constituted almost half of the 2475 observations made during the survey.[24] Several distinct bird communities were identified, including one which consisted of species typically found across southern Australia's mallee belt. For birds such as the southern scrub-robin *Drymodes brunneopygia*, western yellow robin *Eopsaltria griseogularis* and purple-gaped honeyeater *Meliphaga cratitius*, the Yellabinna region represents the northern limit of their distribution in South Australia.

Birds of the Tallaringa region on the desert's eastern edge have recently been studied as part of a major project undertaken by the Australian and New Zealand Scientific Exploration Society.[25]

Factors Shaping Bird Populations

Biogeography

As discussed in the previous section on geology, the Australian continent has undergone a series of major climatic fluctuations during the last 20 million years. A cyclical pattern of humid and wet conditions followed by cold and dry (arid) periods has resulted in the development of concentric zones of habitat types (vegetation formations) comprising distinctive faunal elements. In addition, a series of geographical barriers has formed, causing isolation and subsequent divergence of former continuously distributed taxa.[24,26–29]

Birds of the Great Victoria Desert have affinities with one of three major faunal elements: Torresian (fauna of northern Australia), Bassian (fauna of temperate southern woodlands and forests) and Eyrean (comparatively young fauna of Australia's arid inland).[24] Most are of Eyrean origin, with more than 60 per cent of the 88 species which constitute this faunal element having been recorded in the Great Victoria Desert.[26] They typically represent birds which are widespread across mixed acacia-eucalypt habitats of the central and southern arid zone.

The Nullarbor Plain, located south of the Great Victoria Desert, has acted as a major geographical barrier to the distribution of a number of mainly Bassian birds.[17,24] Examples of species which now have populations in south-western and south-eastern Australia that are separated by the Nullarbor Plain include the restless flycatcher *Myiagra inquieta,* brown-headed honeyeater *Melithreptus brevirostris,* western yellow robin and purple-gaped honeyeater. In most cases, these populations have diverged to a taxonomically recognisable degree. The Nullarbor has also acted as a barrier to the westward spread of a number of introduced species such as the skylark, house sparrow, blackbird and, until recently, the common starling.[17]

A minor but significant geographical barrier within the Great Victoria Desert itself is an area of mallee-covered sand dunes in the Yellabinna region, often referred to in ornithological literature as the Barton Sandhills.[14,28] This belt of country has been implicated in the isolation and subsequent strong divergence within populations of two species in particular, the cinnamon quailthrush *Cinclosoma cinnamomeum* and the blue bonnet *Northiella haematogaster*.

The Naretha blue bonnet Northiella haematogaster narethae *has a restricted distribution, centred on myall woodland on the southern fringe of the Great Victoria Desert.*

Mark Shephard

The Nullarbor form of the cinnamon quailthrush (subspecies *alisteri*) lives on the chenopod shrubland and limestone gibber outcrops of the Nullarbor Plain. The Barton Sandhills separate it from the nominate race, which inhabits similar shrub-steppe and stony gibber country in the Lake Eyre Basin.[26–28] The sandhills also act as a barrier between blue bonnets that inhabit the myall belt fringing the Nullarbor Plain (subspecies *narethae*) and the nominate form which frequents acacia woodlands further east.[19,24,28] Divergence within the disjunct populations of these two bird species has been so strong that morphological and plumage differences have developed. The Nullarbor cinnamon quailthrush has a continuous black throat and breast, while the nominate form has a distinct buff-white band across its upper breast. The Naretha blue bonnet is smaller than its corresponding nominate form and has yellow rather than red on its abdomen. Many ornithologists consider the Nullarbor cinnamon quailthrush and the Naretha blue bonnet as distinct species.

While the Nullarbor Plain and Barton Sandhills have acted as impediments to the movement of certain birds, the Great Victoria Desert has been shown to be a biogeographically active corridor linking mallee-inhabiting species across southern Australia. The relatively unbroken belt of mallee through the Yellabinna region and the southern desert connects mallee communities of birds in the south-western corner of Western Australia with those from Eyre Peninsula in South Australia. As a result, the western and eastern populations of such birds are remarkably uniform and show little or no morphological differences (speciation). Some of the birds which may use this mallee corridor include the purple-crowned lorikeet *Glossopsitta porphyrocephala*, yellow-rumped pardalote *Pardalotus xanthopygus*, Gilbert whistler *Pachycephala inornata*, rufous treecreeper *Climacteris rufa*, grey currawong *Strepera versicolor*, and the chestnut quailthrush *Cinclosoma castanotum*.[12,13,24]

In an interesting quirk of nature, the same system of dunes which splits the two main populations of the cinnamon quailthrush actually connects the western, central and eastern forms of the chestnut quailthrush.[26] The striated grasswren *Amytornis striatus*, while not found in south-western Australia, also uses the desert's mallee belt as a corridor to link Eyre Peninsula populations with those of the Gibson, Great Sandy and Tanami Deserts to the north-west.

The diversity, distribution and abundance of the desert's avian communities is also shaped by other factors, including the variety of habitats the desert supports and the seasonal availability of food and water.

Habitat

The core area occupied by the Great Victoria Desert consists of a mosaic of eucalypt (marble gum woodland and mallee tall shrubland) and acacia (mulga woodland and tall shrubland) communities. The NCSSA and SAOA studies demonstrated that mulga and mallee habitats support a higher density and diversity of bird species than marble gum woodlands. Greater avian diversity in mulga was attributed to an increased number of shrub and ground-dwelling species consequent upon the denser understorey and the range of arid grasses which the mulga woodlands support.[20]

Birds that showed a distinct preference for mulga (over eucalypt) in the NCSSA study included the black-faced woodswallow *Artamus cinereus*, southern whiteface *Aphelocephala leucopsis*, chestnut-rumped thornbill *Acanthiza uropygialis* and little crow *Corvus bennettii*.[19] Those that were more numerous in mulga during the SAOA survey comprised the mulga parrot *Psephotus varius*, red-capped robin *Petroica goodenovii*, white-browed babbler *Pomatostomus superciliosus*, inland thornbill *Acanthiza apicalis* and splendid fairy-wren *Malurus splendens*.[20] Other species found outside the survey areas which are also known to prefer a mulga habitat include the Bourke's parrot *Neophema bourkii* and the chiming wedgebill *Psophodes occidentalis*.

Examples of birds that have been identified as having a decided preference for eucalypt habitat (over acacia-casuarina) include the weebill *Smicrornis brevirostrus*, grey-fronted honeyeater *Lichenostomus plumulus*, striated pardalote *Pardalotus striatus* and red-browed pardalote *Pardalotus rubricatus*. The latter was the only species to show a greater predilection for marble gum woodland over mallee.[19] Other birds that use mallee in the southern desert as a corridor to connect their widely separated populations have been listed previously.

While avian diversity within marble gum communities is generally low, it may well be higher at different times of the year and during favourable seasonal conditions. The large hollow limbs of marble gums undoubtedly provide ideal nesting facilities for parrots when breeding, and roosting sites for year-round nocturnal residents such as owlet-nightjars. Pardalotes and other canopy-feeding insectivorous birds may be more numerous when psyllid (sap-sucking insect) populations are high. Nectivorous species, such as honeyeaters and the purple-crowned lorikeet, would be more common when marble gums are in flower.[20]

The desert's playa lakes support a low, open chenopod shrubland which provides habitats for a different suite of birds, including the white-winged fairy-wren *Malurus leucopterus*, orange chat *Ephthianura aurifrons*, Richard's pipit *Anthus novaeseelandiae*, banded lapwing, and inland dotterel *Peltohyas australis*. These species are also commonly found among stony surfaces and chenopod steppe in the treeless section of the Nullarbor Plain (south of the myall/bluebush belt). Other species associated with the distinctive assemblage of birds from the treeless plain, and therefore not likely to be seen beyond the desert periphery, include the little button-quail *Turnix velox*, brown songlark *Cincloramphus cruralis*, slender-billed thornbill *Acanthiza iredalei*, western calamanthus *Sericornis campestris* and Nullarbor quailthrush.[17]

Seasonal Availability of Food

The influence of seasonal availability of food on the abundance and distribution of the desert's birds is best illustrated by using the honeyeater group as an example.

In 1980, the flowering of common nectar-producing plants such as eucalypts, eremophilas, grevilleas, hakeas and mistletoes was limited in the eastern desert. Honeyeaters constituted only 20 per cent of the total bird count conducted by the NCSSA during the year.[19] Ten years later, when mallee eucalypts (*Eucalyptus socialis, E. striaticalyx, E. youngiana* and *E. kingsmillii*) were flowering profusely, honeyeaters dominated the eastern desert's avifauna.[20] Three species alone (the grey-fronted honeyeater, white-fronted honeyeater *Phylidonyris albifrons* and yellow-throated miner *Manorina flavigula*) accounted for 40 per cent of the total number of bird observations recorded during the 1990 SAOA survey. More than 50 individuals of each of the three species were seen at all sites examined. While honeyeaters followed the flowering mallee in 1990, they were more numerous in mulga woodlands the previous year due to a profusion of flowering eremophilas (*E. gilesii* and *E. latrobei*) associated with the mulga understorey.[20]

Availability of Water

The sole source of fresh water for the desert is natural rainfall which, as stated previously, is very localised and unpredictable. Permanent or semipermanent surface water is limited to scattered rockholes and soaks, although water may transiently pool on the surface of claypans following heavy falls of rain (see Chapter 6).

Lack of surface water has precluded permanent occupation of the desert by aquatic bird species.[19] Seed-eating (granivorous) birds such as pigeons, parrots and finches, which have a daily requirement to drink, have evolved different survival strategies.

The zebra finch Poephila guttata *is one of Australia's most remarkable desert survivors. Adult males display a prominent red cheek patch.*

The scarlet-chested parrot *Neophema splendida* uses dew as a source of water and is able to extract water from grass stems by crushing them in its beak. The zebra finch *Poephila guttata* also uses dew and is capable of drinking highly saline water. Other birds such as the Bourke's parrot and orange chat have the ability to conserve water by lowering their metabolic rate and therefore reducing water loss. The cockatiel *Nymphicus hollandicus* and budgerigar *Melopsittacus undulatus* are examples of highly nomadic parrots which are capable of flying long distances in search of semipermanent sources of water. Their social and reproductive behaviour is fully primed to exploit transient water supplies. Insect-eating birds such as robins, wrens, chats and thornbills, to name a few, and carnivorous raptors, like eagles and falcons, obtain sufficient moisture requirements from their prey.

Desert Species

Many birds are widespread and common throughout the Great Victoria Desert and will almost certainly be seen by the desert traveller (Table 4). In addition, the desert is home to some of Australia's rarest, most enigmatic, yet biologically fascinating species.

Among the birds of prey, the brown falcon *Falco berigora* is widely encountered. Its overall body colour is variable and may range from light sandy brown to very dark brown, almost black. All colour phases feature distinctive dark brown markings below and behind the eye. The brown falcon is often seen roosting on an open branch of a dead tree and may be approached quite closely.

Mark Shephard

Lyn Pedler

ABOVE: The budgerigar Melopsittacus undulatus *is one of Australia's smallest parrots. Its population numbers fluctuate markedly in the Great Victoria Desert, depending on rainfall.*

LEFT: The rufous whistler Pachycephala rufiventris *is a commonly encountered bird in the Great Victoria Desert. The male (pictured) features a deep buff lower breast and abdomen, a white throat, and a black breast band extending around the side of the face and through the eye.*

Lyn Pedler

ABOVE: The hooded robin Melanodryas cucullata *is widespread across the Great Victoria Desert. The male (right) is strikingly marked. The female (left) is mainly grey, but features a white patch on her wings.*

TABLE 4: Examples of common or widespread birds of the Great Victoria Desert[a]

Common name	Scientific name	Common name	Scientific name
Brown falcon	*Falco berigora*	Southern whiteface	*Aphelocephala leucopsis*
Crested pigeon	*Ocyphaps lophotes*	Spiny-cheeked honeyeater	*Acanthogenys rufogularis*
Ringneck parrot	*Barnardius zonarius*	Yellow-throated miner	*Manorina flavigula*
Mulga parrot	*Psephotus varius*	Singing honeyeater	*Lichenostomus virescens*
Red-capped robin	*Petroica goodenovii*	Grey-fronted honeyeater	*Lichenostomus plumulus*
Hooded robin	*Melanodryas cucullata*	White-fronted honeyeater	*Phylidonyris albifrons*
Rufous whistler	*Pachycephala rufiventris*	Mistletoebird	*Dicaeum hirundinaceum*
Grey shrike-thrush	*Colluricincla harmonica*	Black-faced wood-swallow	*Artamus cinereus*
Crested bellbird	*Oreoica gutturalis*		
Willie wagtail	*Rhipidura leucophrys*		
White-browed babbler	*Pomatostomus superciliosus*		
Splendid fairy-wren	*Malurus splendens*		
Variegated fairy-wren	*Malurus lamberti*		
Chestnut-rumped thornbill	*Acanthiza uropygialis*		

[a]Examples are taken from composite bird lists of the author compiled between 1987 and 1994.

The malleefowl is a long-legged, plump, ground-dwelling bird which builds a mound to incubate its eggs. Prior to European settlement, the malleefowl ranged widely over mallee woodland across large areas of inland southern, central and western Australia. Its former distribution and breeding range has declined dramatically during this century due to habitat clearance and burning, reduction of its food plants through grazing by sheep and rabbits, and predation of both young chicks and eggs by foxes in particular. Malleefowl eggs are also a favoured bush food of Aboriginal people. Today, the malleefowl is moderately abundant in remnant pockets of mallee. Its status in South Australia is considered endangered, while throughout Australia as a whole it is classed as vulnerable.[24,30]

The malleefowl is presumed to be sparsely distributed in the Great Victoria Desert. Active mounds have been recorded over the past 25 years from near the Serpentine Lakes, east of Vokes Hill Corner, north-west of Maralinga, near Cheesman Peak in the Anangu Pitjantjatjara Lands, south-west of Mount Christie railway siding, and in the Yumbarra Conservation Park adjoining the southern tip of the Yellabinna region.[31] Recent breeding records have come from Ilkulka (between Neale Junction and the Serpentine Lakes), Queen Victoria Spring Nature Reserve and east of Lake Minigwal.[32]

The mound-building activities of the malleefowl are quite remarkable. Both male and female birds (who form a mated pair for life) scrape a mound from leaf litter, twigs, bark and sand, using their strong legs and powerful claws to great effect. Once completed, the mound may be up to 5 metres in diameter and 1.5 metres high. The female then spends most of her time searching locally for food and only returns to the mound to lay eggs. In good seasons and at productive sites, around 30 eggs may be deposited in an egg chamber which is dug into the top of the mound and covered with sand. The eggs are incubated by the heat generated through the fermentation of rotting leaf litter and by the sun directly warming the sand. The male maintains and defends the nest throughout the year. It is his responsibility to ensure that incubation temperature in the mound is constantly around 33°C. Temperature is checked by driving his open bill into the mound, after which minor adjustments are made to the mound by either adding or removing material. When the chicks hatch, they dig their own way out of the mound and are able to fly within 24 hours of surfacing.

The Australian bustard *Ardeotis australis* is another plump game bird which invariably displays a stately, erect posture. Habitat clearance, predation by foxes and cats, and traditional hunting by Aboriginal people have reduced the range of this bird, which was once a common nomad across southern Australia. In the Great Victoria Desert, it is most likely to be seen in open grassland among woodland (especially in recently burnt areas).

The crested pigeon *Ocyphaps lophotes* is a member of the bronzewing family of pigeons which show iridescent patches in their wing feathers. This stocky brownish-grey bird with a red eye-ring and an erect, slender black crest, is common throughout the desert. It flies with a characteristic 'whistle of wings' and cocks its tail on alighting.

The common bronzewing *Phaps chalcoptera* is more sparsely distributed and less commonly seen than the crested pigeon. This ground-feeding bird has a delightful russet-cream forehead and a white stripe beneath and behind the eye. It displays distinct scalloping and conspicuous bronzing over its wing coverts. A group of eight common bronzewings has been recorded drinking at dusk at Tjintirrgarra waterhole, just west of the Serpentine Lakes.[19]

The mulga parrot *Psephotus varius* and ringneck parrot *Barnardius zonarius* are widely distributed and regularly encountered throughout the desert. The male mulga parrot is brilliant emerald green overall but features a striking array of other bright colours including yellow on the frontal band, shoulder and undertail; red on the rump, thighs, abdomen and behind the crown; and blue on the fold of the wing and tail. It is often aptly referred to as the many-coloured parrot. The female mulga parrot is much duller, being brownish-green overall with a red shoulder patch.

There are four major forms of the ringneck parrot which range widely from the south-western corner of Western Australia, across the arid interior of southern Australia, and east to New South Wales and the Cloncurry area of Queensland. All forms intergrade or hybridise where their ranges meet. The ringneck parrot found in the Great Victoria Desert is generally referred to as the Port Lincoln parrot. It is a large, broad-tailed green parrot which has a black head, yellow abdomen and distinctive yellow ring or collar on its hindneck.

Several other parrots occur patchily across the desert. The delightful Bourke's parrot *Neophema bourkii*, with its soft blue, grey and pink plumage, has been regularly recorded in the Tallaringa Conservation Park on the desert's eastern edge where mulga predominates.[33] Vokes Hill appears to be the western limit of its distribution on the South Australian side of the Great Victoria Desert.[34] Given its distinct preference for mulga habitat, some ornithologists argue that the Bourke's parrot is not truly a bird of the Great Victoria Desert, but rather a peripheral species.[34]

The Naretha blue bonnet is largely confined to the myall woodland west and north-west of the Nullarbor Plain. It is widespread and moderately common within this area.[14] It is less frequently observed south of the Transcontinental Railway Line. In South Australia, a small population of fewer than 200 birds has been recorded in a broad belt of myall north of Colona and south of Ooldea. The existence of this population may be due in part to the release of illegally trapped birds, although certainly the habitat exists in this area to sustain a naturally occurring population.[30,34] The Naretha blue bonnet has been recorded nesting in myall, mulga, myoporum and black oak. In the latter, the nest entrance was through cracks or splits that had developed around the base of live or dead trees.[14]

The pink cockatoo *Cacatua leadbeateri*, with its majestic red and orange crest, beautiful salmon-pink plumage and powerful yet graceful flight, is one of Australia's most attractive birds. It is classed as a vulnerable species in South Australia. While only sparsely distributed in the Great Victoria Desert (usually in pairs or small flocks), the general inaccessibility of the desert is considered important to the species' conservation.[24] The pink cockatoo has been observed feeding on ripe fruits of the rattlepod grevillea and tearing the branches of the desert poplar while searching for wood-boring grubs.[20] It favours hollows of the marble gum for nesting, although it has been observed breeding in large black oaks in the northern part of the Yellabinna region.[35]

Two other 'crowning jewels' of the parrot family are found in the Great Victoria Desert: the scarlet-chested parrot and the Alexandra's or princess parrot *Polytelis alexandrae*.

The scarlet-chested parrot is unquestionably one of Australia's most beautiful birds. The adult male displays a visually stunning combination of bright colours: a brilliant blue face, scarlet chest, buttercup yellow abdomen and bright green upperparts.

Although the range of the scarlet-chested parrot extends over large tracts of arid southern Australia, its status is generally considered rare.[30] The Great Victoria Desert, and in particular the Yellabinna region, is now regarded as the centre of distribution of the species.[19,24] Historically, a transient 'irruption' of scarlet-chested parrots occurred on the southern periphery of the Great Victoria Desert at Wynbring railway siding in 1939. More than 100 birds, as well as eight nesting sites (mainly in marble gum hollows), were recorded in the centre of the desert between May 1968 and 1970.[13] Extensive field work in the Great Victoria Desert since 1970 has found only small numbers of scarlet-chested parrots, with occasional breeding records being documented.[19,24,31,36–37]

During the winter of 1993, the largest flock of scarlet-chested parrots seen in the Great Victoria Desert for 25 years was recorded independently by myself and by a party of ornithologists from the RAOU.[38,39] Around 240 birds were observed by the RAOU team on 13–15 May at a location 9 kilometres south of Vokes Hill Corner in the Unnamed Conservation Park. A minimum of 120 birds, presumably the same flock, were sighted in the same general area by my party on 22–23 June. The birds were remarkably tame and could be approached to within a few metres. The ratio of males to females in the flock was around 2:1. Many males showed only patchy red colouration on the chest, indicating they were juvenile birds around six months of age. The flock presumably represented a loose, post-breeding congregation of birds.

Lynton Huxley

Lynton Huxley

LEFT: The author recorded a large flock of scarlet-chested parrots Neophema splendida *near Vokes Hill Corner during the winter of 1993. The adult male (pictured) is a visually outstanding bird.*

ABOVE RIGHT: The status of Alexandra's parrot Polytelis alexandrae *in the Great Victoria Desert is unknown. The centre of its distribution now appears to be the sandridges of the Great Sandy Desert.*

BELOW: Alexandra's parrot has recently been recorded between Vokes Hill Corner and the Serpentine Lakes along the Anne Beadell Highway.

Mark Shephard

Habitat in which the birds were found comprised mixed mallee and acacia tall shrubland, with a spinifex (*Triodia*) and tussock grass understorey. Marble gum was conspicuous by its absence. Birds were observed feeding on desert thryptomene which was seeding prolifically in the area.

The large number of juvenile birds seen near Vokes Hill probably resulted from an opportunistic breeding response triggered by heavy local rainfall over the early summer of 1992. The finding of a young scarlet-chested parrot chick in the Yellabinna area in early May 1992 provides supportive evidence that nesting of the scarlet-chested parrot may occur at any time of the year in response to favourable conditions.[31]

Further observations on the scarlet-chested parrot were made by my party in the winter of 1993. Four birds (two adult pairs) were seen feeding in a sandy depression ringed by 5-metre sandridges, approximately 100 kilometres west of Vokes Hill Corner. The depression contained only three plant species: seeding *Triodia*, desert thryptomene and flowering Latrobe's desert fuschia. The vegetation of the sandridges and surrounding area comprised marble gum, mallee, mulga and black oak.

Alexandra's parrot, or the princess parrot as it is commonly known, is unquestionably one of Australia's most alluring, yet enigmatic birds. Its range is thought to extend over most of arid western and southern Australia and the Northern Territory, but sporadic sightings of the bird together with its purported nomadic nature make its status difficult to assess. It is presumed to be at least rare.[30]

The Alexandra's parrot's exquisite pastel shades of blue and green, rose-coloured throat, and long, slender tail combine to make it one of the most highly sought-after species amongst Australian ornithologists. For many enthusiasts (including myself), it is a life-long ambition to see this bird in its natural environment.

In the spring of 1965, a swift-flying flock of around 40 Alexandra's parrots was sighted in mulga on the Connie Sue Highway around 115 kilometres north of Neale Junction. A second group of 12 birds was later recorded 70 kilometres east of Cosmo Newbery.[12]

The first sighting of Alexandra's parrot on the South Australian side of the Great Victoria Desert was made as recently as September 1986.[40] (This sighting represented only the fifth record of the bird in South Australia as a whole.) Four birds were seen 24 kilometres west of Vokes Hill Corner in the Unnamed Conservation Park by Ross Allen, District Ranger for the Far West Region of the National Parks and Wildlife Service of South Australia. The habitat at this location comprised mulga, black oak and marble gum, with an understorey of senna, eremophila and *Triodia*. Birds appeared to be feeding on the seeds of *Stenopetalum anfractum*.[40]

In October 1989, George and Ruth Aspley recorded four Alexandra's parrots, 136 kilometres west of Vokes Hill Corner.[41] The birds initially flew alongside one of the party's lead vehicles. They were followed by the vehicle until they flew into a marble gum and were later observed as they fed on the ground near *Triodia* and roosted in mulga trees.[41]

In October 1993, the Alexandra's parrot was also sighted in the Tallaringa Conservation Park on the far eastern edge of the Great Victoria Desert during an ANZSES expedition to the region.[25]

Recent observations of large numbers of Alexandra's parrots along the Canning Stock Route (including my own observations of around 100 birds between Wells 38 and 42) have initiated new theories on the ecology and life history of the parrot. It appears that the sandridges of the northern Great Sandy Desert may represent a core habitat for Alexandra's parrot.[42] Birds work the sand dune country, setting up territories along dunes where they feed and breed in good seasons. When the core area is in drought, they move away from the sandridges and concentrate around selected rockholes in range country. The birds become more nomadic in prolonged drought and move eastwards (to the western MacDonnell Ranges) and southwards (to the Great Victoria Desert).[43]

The red-backed kingfisher *Halcyon pyrrhopygia* lives among the desert's open mulga and mallee woodlands, where it is often seen perched motionless on a dead branch. It is nomadic over much of Australia's arid interior and is mostly carnivorous in its dietary habits. The red-backed kingfisher has a distinctive streaked crown and broad black band running from behind the eye to the back of the neck. It has light turquoise blue wings, a rufous rump and uppertail covets, and white underparts.

Two members of the robin family are commonly seen in the desert. They are the red-capped robin *Petroica goodenovii* and the hooded robin *Melanodryas cucullata*. The colourful males of both species are generally black on the back and have a prominent white wing bar, but may be readily distinguished by major differences in breast and head plumage. The red-capped robin has a delightful scarlet breast and cap, while the hooded robin has a completely black head and upper breast and white underparts. Robins are insectivorous and build a neat, cup-shaped nest, usually in the fork of a tree. The nest may be decorated or lined with lichen, moss or spider webs.

The crested bellbird *Oreoica gutturalis* is one of the true characters of Australia's dry interior. Its repetitive, ventriloquial call, which has been likened to morse code by well-known outback personality Len Beadell, will be heard constantly by the traveller to the Great Victoria Desert. Although its call is never far away, sighting the bird can prove more difficult and frustrating. The grey-brown male has a distinctive black breast band and crown, white around the forehead and throat, and buff underparts. Its crown feathers may be raised to a small crest. The female lacks the crest, black breast band and white face, but retains the narrow strip of black feathers extending from the crown to the nape.

The white-browed babbler is another common bird of the Great Victoria Desert. It lives in highly sociable, mainly sedentary groups of around a dozen birds. The group is very vocal and constantly chattering, as its members 'bounce' along the ground or fly low from bush to bush. Numerous domed stick nests are built in tree forks or horizontal branches and are used both for breeding and communal roosting. The white-browed babbler is omnivorous and takes most of its food from the ground. It is readily identified in the field by its dark crown, prominent white eyebrow, and white breast merging into a grey-brown abdomen.

Fairy-wrens are one of Australia's most delightful families of birds. They live in small cooperative family groups of up to eight birds, consisting of a dominant male, one adult female and subordinate non-breeding birds. The latter assist in raising young chicks, a behaviour which enables the breeding female to raise up to three broods in a season. The breeding male is always strikingly coloured, but undergoes a moult into an eclipse plumage after the breeding season until around four years of age. Three species of fairy-wren occur in the Great Victoria Desert, namely, the white-winged fairy-wren *Malurus leucopterus*, the variegated fairy-wren *M. lamberti* and the splendid fairy-wren *M. splendens*. The white-winged fairy-wren has been recorded among low chenopod shrubland (including sandridges) around the Serpentine Lakes.[34] The variegated and splendid fairy-wrens are more widespread through the Great Victoria Desert's mulga and mallee country. The male splendid fairy-wren is perhaps the most spectacular of all the fairy-wrens. His whole body is iridescent cobalt blue which 'glows' in full sunlight. The female is dull brown, but has a bright blue tail and a chestnut patch around the eye.

Thornbills are small insectivorous birds which feed in small groups and build dome-shaped nests. Throughout Australia there are around 12 species of thornbills. Each may be distinguished by facial and chest markings, eye colour, rump colour and call. Two species commonly found in the Great Victoria Desert are the inland thornbill *Acanthiza apicalis* and the chestnut-rumped thornbill *A. uropygialis*. The inland thornbill has a red eye, grey-brown scalloped forehead with white crescents, and a streaked breast. The chestnut-rumped thornbill has a white eye, light brown forehead lightly speckled with white, and a plain breast.

The weebill is Australia's smallest bird (total length: 8 cm). It is similar to thornbills in many respects but has a shorter, stouter bill. It will be found almost exclusively in, or flying between, the canopies of eucalypts.

The southern whiteface is relatively common in open areas of mulga with a grassland or low shrub understorey. It is grey-brown and cream to white below, and has a distinctive white forehead and face which is separated from the crown by a dusky line in front of its eyes. The southern whiteface forages on open ground, often in small parties. If disturbed, it may fly to a dead branch and constantly twitter in alarm.

The brush-tongued honeyeater family is one of Australia's most successful bird groups. Honeyeaters feed on a variety of foods including nectar, fruit, insects and sugary secretions of herbivorous insects such as lerps and honeydew. The diversity and abundance of honeyeaters in the Great Victoria Desert is dependent to a large extent on seasonal availability of food. However, several species are always readily seen.

The grey-fronted honeyeater is found mainly in eucalypt woodlands and areas where Latrobe's desert fuschia is flowering. It has a distinctive yellow plume and buff-grey underparts lightly streaked with brown. The common name, grey-fronted honeyeater, is somewhat confusing as grey is restricted to the bird's frons or forehead.

The yellow-throated miner is a grey-brown bird which has a yellow bill and two yellow streaks of bare skin extending from the bill. Patches of yellow are also featured on the forehead, eye and side of the neck. It is a highly sociable bird and is usually seen in small vocal flocks.

The spiny-cheeked honeyeater is a small wattlebird which has distinctive white 'spiny' feathers from its bicoloured bill to its cheek. Its throat and breast are cinnamon and its abdomen is creamy white and spotted with brown. It also has a white rump which is easily seen during flight.

The white-fronted honeyeater is a yellow-winged honeyeater with a black throat and breast, lightly streaked white abdomen, and a small red wattle behind the eye. Again the name 'white-fronted' is somewhat misleading as white is restricted to the forehead, lores, eyering and moustache around the facial area.

Lyn Pedler

The malleefowl Leipoa ocellata *is sparsely distributed throughout the Great Victoria Desert. Malleefowl eggs are a favoured traditional food of Aboriginal people.*

The singing honeyeater *Lichenostomus virescens* is grey-brown in colour and has a distinctive black band of feathers running from the bill through the eye and onto the neck. There is a yellow streak (malar plume) below this band which ends in a white tuft behind the ear. Its breast and underparts are pale grey and faintly streaked with dark grey.

The black honeyeater and the pied honeyeater are two long-billed, black and white honeyeaters that frequent the desert in good seasons. The black honeyeater has a black stripe down the centre of its belly, while the pied honeyeater displays a white wing patch.

The yellow-plumed honeyeater is officially listed as vulnerable in South Australia and is essentially a bird of the mallee. Despite its status, it is widespread and common in the Yellabinna region which appears to be a critical area for its survival. In the far western part of Yellabinna, its range overlaps with the mallee-loving grey-fronted honeyeater which is found widely throughout the Great Victoria Desert. The yellow-plumed honeyeater has also recently been recorded in the Unnamed Conservation Park.[35]

Pardalotes are small, colourful birds with short tails and stout blunt beaks. They forage throughout the canopy of eucalypts, gleaning its foliage for sugary lerps. The striated pardalote is widespread in the desert's marble gum and mallee communities. It has a black-and-white streaked crown, yellow eyebrow and throat, and white wing stripe with a red spot. It gives a double or triple-noted call. The red-browed pardalote features an orange-red eyebrow, dark crown spotted with white, yellow wash on its breast, and yellow edges to its wing feathers. It has a five-noted call, with each note increasing in pitch and speed. The red-browed form is less common than its striated counterpart and prefers marble gum habitat to mallee. The yellow-rumped pardalote, listed as vulnerable in South Australia, has been recorded in the southern and western edge of the Yellabinna region.

The black-faced woodswallow, which is smoky grey in colour, is one of the most common birds of the Great Victoria Desert. The masked woodswallow *A. personatus* is much paler grey and has a more extensive area of black around the face and throat. It may also be frequently encountered in the desert. The little woodswallow *A. minor* is a small chocolate-brown bird. It is an occasional visitor from the central ranges and is generally restricted to cliff areas adjacent to salt lakes. The dusky woodswallow *A. cyanopterus,* which is smoky brown and has a distinctive white edge to its wings, has only been recorded from the southern edge of the Great Victoria Desert.

Chapter 4

MAMMALS

The Desert Mammals

Historical information on the mammals of the Great Victoria Desert is available from several sources. The accounts of early explorers to the region, notably Richard Maurice and the Elder Expedition (see Chapter 7), provide a valuable insight into the desert's mammal fauna at the turn of the century.[1]

Prominent zoologists such as Frederick Wood Jones and Ellis Troughton made occasional observations of mammals in the Ooldea region during the 1920s and 1930s.[2] Anthony Bolam and Daisy Bates also collected many specimens of mammals for the South Australian, Western Australian and Australian Museums from Ooldea around the same period.

Bolam, who was stationmaster at Ooldea from 1918 to 1925, wrote up his observations on the flora and fauna of Ooldea in a delightful book entitled *Trans-Australian Wonderland*.[3] His records were remarkably detailed and perceptive for an amateur naturalist. Bolam's style of writing reflected his deep passion and interest in things natural and his book makes enchanting reading. Bates also wrote extensively on the natural history of the region, basing her material mainly on information solicited from Aboriginal people who lived at or visited Ooldea. She frequently corresponded with both Frederick Wood Jones and Hedley Finlayson.[1]

The early mammal records from Ooldea need to be interpreted cautiously, however. Ooldea sits at the interface between two vastly different arid landscapes, the Great Victoria Desert and the Nullarbor Plain. In some instances, it is unclear from which of these two regions specimens were taken. Other specimens and information may have come from Aboriginal people who travelled from well beyond the boundaries of the Great Victoria Desert to reach Ooldea (see Chapter 6).[2,4]

Little work was published on the mammals of the Great Victoria Desert during the ensuing 50 years. A flurry of scientific activity has occurred since the mid-1970s, with surveys being conducted in proposed and existing Western Australian nature reserves, the Unnamed Conservation Park, the Yellabinna and Tallaringa regions, and the neighbouring Nullarbor Plain.[2,4–9] A variety of techniques are now used in modern-day research to obtain information on mammal fauna. These include Elliott and pitfall traps for ground mammals, mistnets and harp traps for bats, spotlighting, and examination of skeletal remains, scats (faecal pellets) and tracks.[10]

TABLE 5: The present-day native mammals of the Great Victoria Desert regions[a]

Family	Common name	Scientific name	Unnamed CP [b]	Yellabinna Wilderness	Tallaringa CP	Queen Victoria Spring NR	Plumridge Lakes NR	Neale Junction NR	Yeo Lake NR
Tachyglossidae	Short-beaked echidna	*Tachyglossus aculeatus*		+	+				+
Dasyuridae	Mulgara	*Dasycercus cristicauda*				+			
	Kultarr	*Antechinomys laniger*		+					
		Sminthopsis dolichura		+		+			
	Ooldea dunnart	*Sminthopsis ooldea*	+	+	+	+		+	
	Fat-tailed dunnart	*Sminthopsis crassicaudata*	+	+	+				
	Hairy-footed dunnart	*Sminthopsis hirtipes*	+	+		+		+	
	Stripe-faced dunnart	*Sminthopsis macroura*			+				
	Sandhill dunnart	*Sminthopsis psammophila*		+		+			
	Wongai ningaui	*Ningaui ridei*	+	+		+		+	
		Ningaui yvonneae		+					
	Marsupial mole [c]	*Notoryctes typhlops*							
Burramyidae	Western pygmy possum	*Cercartetus concinnus*		+		+			
Macropodidae	Western grey kangaroo	*Macropus fuliginosus*	+	+		+			
	Euro	*Macropus robustus*		+		+			+
	Red kangaroo	*Macropus rufus*	+	+	+		+		+
Muridae	Sandy inland mouse	*Pseudomys hermannsburgensis*	+	+	+	+	+	+	+
	Bolam's mouse	*Pseudomys bolami*					+		
	Mitchell's hopping-mouse	*Notomys mitchelli*	+	+		+			
	Spinifex hopping-mouse	*Notomys alexis*	+		+	+	+	+	+
Vespertilionidae	White-striped mastiff-bat	*Tadarida australis*		+		+			
	Little mastiff-bat	*Mormopterus planiceps*				+			
	Greater long-eared bat	*Nyctophilus timoriensis*	+	+					
	Lesser long-eared bat	*Nyctophilus geoffroyi*	+	+	+	+			
	Gould's wattled bat	*Chalinolobus gouldii*			+	+			
	Western broad-nosed bat	*Scotorepens balstoni*			+				
	Little broad-nosed bat	*Scotorepens greyii*			+	+			
	Inland eptesicus	*Vespadelus baverstocki*		+					
	Little cave eptesicus	*Vespadelus finlaysoni*				+			+

[a]Reference source for each region: Unnamed Conservation Park (reference 7), Yellabinna (4), Tallaringa (8,9), Queen Victoria Spring, Plumridge Lakes and Neale Junction Nature Reserves (5) and Yeo Lake (6). Information on mammals of Queen Victoria Spring Nature Reserve was also obtained from D. Pearson (pers. comm.).

[b]Species listed under the Unnamed Conservation Park also include those recorded along the Vokes Hill Corner to Cook road.[7]

[c]The marsupial mole has been recorded in the Great Victoria Desert Nature Reserve.[5]

NB: As research continues, it is likely that more species will be found occurring in the Great Victoria Desert. The fat-tailed antechinus *Antechinus macdonnellensis*, for example, is likely to occur in the desert's rocky outcrop and breakaway country.

Peter Canty

ABOVE: The sandhill dunnart Sminthopsis psammophila *has recently been recorded in the Queen Victoria Spring Nature Reserve and the Yellabinna region.*

RIGHT: The Ooldea dunnart Sminthopsis ooldea *was first collected from the tiny railway siding of Ooldea. It remained virtually unknown until the 1970s, but with the advent of pitfall traps it is now regarded as widely distributed in the Great Victoria Desert.*

Peter Canty

Present-Day Native Mammals

Based on the results of these recent surveys, the desert's present-day native mammal fauna is thought to comprise around 30 species (Table 5).

Most of the native ground mammals inhabiting the Great Victoria Desert are those which prefer a sandy habitat, or those which occupy a broad range of habitats including sand. Breakaway country in the north-western desert, salt lakes and clay-pans, and stony surfaces provide alternative habitats for other species not found in the desert's sandy core.

Egg-laying mammals (monotremes) are represented in the Great Victoria Desert by the short-beaked echidna *Tachyglossus aculeatus*. The echidna is well known for its sharp, spiny coat and long tongue which it uses to great effect in devouring ants and termites. It has been recorded in areas as widely separated as Yellabinna and Tallaringa in South Australia, and the Yeo Lake region in Western Australia.[4,6,8] The echidna generally prefers more rocky habitats or sandplains and sand dunes adjacent to rocky areas. It is still widely hunted by Aboriginal people for its excellent meat.[11]

Small carnivorous marsupials, known as dasyurids, are well represented in the Great Victoria Desert due to their preference for hummock grassland (*Triodia*) habitats. Different-sized dasyurids were shown to coexist at the same survey site in the Yellabinna study.[4]

The mulgara *Dasycercus cristicauda*, which features a characteristic crest of short black hairs on its tail, is known to inhabit sandy regions of arid Australia. Bolam recorded and collected the mulgara in the sandhills around Ooldea in the 1920s.[3] He described its voracious appetite as follows:

> *On one occasion, a young phascogale [mulgara], about 6 inches [15 cm] long was placed in a cage beside a full-grown rabbit. To our great surprise on the follow-up morning, we found the rabbit dead and the young phascogale asleep upon its body.*[3]

A single female mulgara was captured in the Queen Victoria Spring Nature Reserve near the south-west corner of the desert in December 1987.[12] This specimen represents a first record for the Reserve. The site, 25 kilometres north-north-east of Queen Victoria Spring, was dominated by thick hummock grassland (*Triodia basedowii*), marble gum, Ooldea mallee and a diverse shrub layer. It is likely that the mulgara still occurs in scattered populations in other parts of the Great Victoria Desert.

The kultarr *Antechinomys laniger* was once common over much of Central Australia. The only recent record of this marsupial in the Great Victoria Desert is a single specimen collected during the Yellabinna survey.[4] It was captured in a sandplain habitat comprising mixed shrub, mulga and mallee.

The dunnarts are another interesting group of dasyurids; they are represented in the Great Victoria Desert by the sandhill dunnart, hairy-footed dunnart, fat-tailed dunnart, stripe-faced dunnart, Ooldea dunnart and long-tailed dunnart, as well as *Sminthopsis dolichura*.

The greyish-brown Ooldea dunnart *Sminthopsis ooldea* was first collected from Ooldea, as its scientific name suggests. It now appears to be the most widely distributed dunnart of the Great Victoria Desert, inhabiting acacia and mallee scrubland and hummock grassland. It is considered to be nocturnal and insectivorous.

Until 25 years ago, the sandhill dunnart *Sminthopsis psammophila* was known only from an adult male captured near Lake Amadeus in Central Australia during the 1894 Horn Expedition. However, two specimens were collected from the south-western corner of the Great Victoria Desert in the mid-1980s.[13] A further five individuals (two males and three females) were captured within a 1.5 by 0.5 kilometre area of sandplain on the northern boundary of the Queen Victoria Spring Nature Reserve between June 1987 and March 1988.[14] In the 1987 Yellabinna survey, five sandhill dunnarts (three males and two females) were also recorded at three sites, 150 kilometres apart: Yarle Lakes, Ooldea and Mount Christie.[4,14] The habitats in which all 12 dunnarts were caught were similar, and consisted of a relatively diverse shrub layer and a spinifex understorey with a ground cover of more than 20 per cent. Soils were sandy in texture. The sandhill dunnart is the largest of all dunnarts and is classed as rare in both South Australia and Western Australia.[4,15]

The hairy-footed dunnart *Sminthopsis hirtipes* is so-named because of the fine bristles that cover the pads of its feet. Until recently the species was considered rare. New records from west of Vokes Hill in the Unnamed Conservation Park, Muckera, Queen Victoria Spring Nature Reserve, north of Neale Junction, and the De La Poer Range Nature Reserve suggest that its status is now secure.[4,5,7,16] Most of these latest records are from sandy areas among open marble gum woodlands.

The remaining dunnarts of the Great Victoria Desert appear more restricted in their range. Recent records of the fat-tailed dunnart *Sminthopsis crassicaudata* in the eastern Great Victoria Desert come from the Yellabinna and Tallaringa regions and the Unnamed Conservation Park.[4,7–8] The stripe-faced dunnart *S. macroura* has been collected from the Tallaringa area and the De La Poer Range Nature Reserve at the eastern and western extremities of the desert respectively.[8,16] *S. dolichura* was common in mallee-covered dunes during the Yellabinna survey.[4]

Ningauis are very small, aggressive, nocturnal marsupials.[15] Wongai ningaui *Ningaui ridei* has been captured in pit traps set among marble gum woodlands/dunefield habitats in both the Queen Victoria Spring and Neale Junction Nature Reserves during the late 1970s.[5] The single animals collected at each location represented only the third and fourth specimen of the species to be collected respectively. Six further specimens were captured among spinifex grassland in dunefields and sandplains south of Wyola Lakes during 1980.[7] Both *Ningaui ridei* and *Ningaui yvonneae* were recorded in the 1987 Yellabinna Survey.[4]

The marsupial mole *Notoryctes typhlops* is a remarkable desert mammal which spends most of its life underground, occasionally coming to the surface after heavy rain. It has no external ears, is totally blind, and burrows through the sand by making use of flattened spade-like claws and a horny shield which protects its snout. Bolam commented that workmen came across marsupial moles on several occasions during the excavation of railway cuttings around Ooldea. He made the following comical observation of the mole in captivity:

> *...it will make tours of its cage; round and round it will go until suddenly it will drop off to sleep quite surprisingly, and it will just as suddenly wake up, and off it will rush again. It will even fall asleep in the middle of a meal, wake up with a start, and with the utmost speed devour what is left.*[3]

The marsupial mole has been recently recorded in the Great Victoria Desert Nature Reserve.[5] It was also reported on the north-western edge of the desert by Jan Turner, anthropologist with the Ngaanyatjarra Council, in 1991.

The western pygmy possum *Cercartetus concinnus* is the only living representative of the possum families in the Great Victoria Desert. It has been recorded in the mallee woodlands of the Yellabinna region and in the Queen Victoria Spring Nature Reserve.[4] This tiny ball of soft fur uses its long prehensile tail to great effect in climbing through trees. It spends the day asleep inside tree hollows before waking at night to feed on insects and nectar.

Kangaroos found in and around the desert include the red kangaroo *Macropus rufus*, western grey kangaroo *M. fuliginosus* and euro *M. robustus*. The western grey kangaroo has been documented sporadically along the southern edge of the desert in the Yellabinna, Queen Victoria Spring and Nullarbor Plain areas. Recent sightings west of Vokes Hill and in spinifex-covered sand dunes north to Neale Junction have extended its known range well into the desert.[7,16]

The red kangaroo is Australia's largest living marsupial. It prefers to graze on tussock grasses among the desert's open plains or open mulga and eucalypt woodlands, but is generally absent from sandplains and dunefields where spinifex predominates.[11]

The smaller euro has a distinct habitat preference for rocky hills or stony rises. It is therefore most likely to be found in breakaways along the desert's north-western edge and around the Mt Finke area in the south-eastern corner of Yellabinna.[4]

The rodent family is poorly represented in the Great Victoria Desert. However, three species are widespread throughout much of the desert: the Mitchell's hopping-mouse *Notomys mitchelli*, spinifex hopping-mouse *Notomys alexis* and sandy inland mouse *Pseudomys hermannsburgensis*. The numbers of each species may fluctuate according to prevailing climatic conditions, but their high reproductive rate and early attainment of sexual maturity ensure that favourable seasons and the resultant abundant food supply are fully exploited.

Peter Canty

Peter Canty

ABOVE: *The fat-tailed dunnart* Sminthopsis crassicaudata *is widespread across arid southern Australia. Fat stores in its swollen tail function as an energy reserve.*

RIGHT: *Wongai ningaui* Ningaui ridei *is a small, aggressive marsupial, which is only 6 cm in length. It is pictured feeding on a dead military dragon* Ctenophorus isolepis.

Hopping-mice are delightful and highly social rodents with long hind legs and tufted tails. They shelter during the day in burrows which provide a cool retreat from the desert sun.[15] Bolam described their burrows as 'fine examples of strategic construction which our military engineers might copy to advantage'.[3] They feed after dark on seeds, insects, green plants and roots. Mitchell's hopping-mouse is the largest living member of this group. Its distribution appears to be centred more on mallee areas in the southern part of the Great Victoria Desert, although both Mitchell's hopping-mouse and the spinifex hopping-mouse were collected along the same trap line near the Vokes Hill area.[7] Mitchell's hopping-mouse was the most commonly recorded mammal (117 individuals) in the Yellabinna survey.[4] The spinifex hopping-mouse appears to be more wide-ranging across the whole desert. It is well known for its ability to exist for long periods without free water by recycling metabolic water and concentrating its urine.

The sandy inland mouse is found over much of arid central and Western Australia, often living together with the spinifex hopping-mouse.[15] It shares similar habitats with hopping-mice and is perfectly adapted to desert life among sandhills and mulga scrub. Bolam's mouse *Pseudomys bolami*, a close relative of the sandy inland mouse, is common in southern parts of the desert and prefers harder soils.

A number of insectivorous bats may be found in the Great Victoria Desert. Bats are the only mammals capable of flight and have the ability to navigate and find prey by emitting an ultrasonic, sound-reflecting call (echolocation). Many of the bats found in the Great Victoria Desert are species that range widely across the Australian continent, for example Gould's wattled bat *Chalinolobus gouldii*, the lesser long-eared bat *Nyctophilus geoffroyi*, little broad-nosed bat *Scotorepens greyii* and the western broad-nosed bat *Scotorepens balstoni*. Some are found mainly in the southern half of the continent (white-striped mastiff-bat *Tadarida australis*, little mastiff-bat *Mormopterus planiceps* and inland eptesicus *Vespadelus baverstocki*). Others have more restricted or patchy distributions which include the Great Victoria Desert (for example, the greater long-eared bat *Nyctophilus timoriensis*).

Tale of Extinction

The preceding section reveals only part of the story of the desert's native mammals. By the time modern-day research commenced, many of the mammals originally recorded around the turn of the century and in the 1920s and 1930s had vanished. It is now known that the mammal fauna of the Great Victoria Desert (and the Central Australian desert region as a whole) has suffered a dramatic decline virtually 'under the cover of darkness' during the past 50 to 80 years. This demise has been most pronounced among the desert's medium-sized mammals (weight: 35–5500 grams).[11]

RIGHT: The marsupial mole Notoryctes typhlops *burrows into the desert sands using its flattened claws and nose. The only signs of its presence are its tracks and a small crater of sand at the point of underground entry.*

BELOW: The sandy inland mouse Pseudomys hermannsburgensis *is a highly social rodent which shelters in deep burrows during the day and feeds at night on seeds, grasses and roots.*

Peter Canty

Peter Canty

TABLE 6: Some native mammals now presumed to be extinct throughout most or all of the Great Victoria Desert

Family	Common name	Scientific name
Dasyuridae	Western quoll	*Dasyurus geoffroii*
Myrmecobiidae	Numbat	*Myrmecobius fasciatus*
Peramelidae	Pig-footed bandicoot	*Chaeropus ecaudatus*
	Western-barred bandicoot	*Perameles bougainville*
	Desert bandicoot	*Perameles eremiana*
	Golden bandicoot	*Isoodon auratus*
	Greater bilby	*Macrotis lagotis*
Phalangeridae	Common brushtail possum	*Trichosurus vulpecula*
Potoroidae	Brush-tailed bettong	*Bettongia penicillata*
	Burrowing bettong	*Bettongia lesueur*
Macropodidae	Rufous hare-wallaby	*Lagorchestes hirsutus*
	Crescent nail-tail wallaby	*Onychogalea lunata*
Muridae	Lesser stick-nest rat	*Leporillus apicalis*

A closer scrutiny of historical records and the soliciting of traditional ecological knowledge of Aboriginal people have both helped to 'fill in the gaps' regarding what happened to many of the desert's mammals and when they vanished.[1,11,18–20] An important study documenting Aboriginal knowledge of mammals from the central deserts region, including the Great Victoria Desert, was carried out recently by scientists from Western Australia and the Northern Territory.[11] Aboriginal people were shown museum skins of mammals and asked to provide information on local names, past and present status, and aspects of biology and ecology. This study provided new information on the timing of extinction of many species and extended the former range of others.

At least 13 mammals are now thought to have become extinct in the Great Victoria Desert since European settlement (Table 6).

The numbat *Myrmecobius fasciatus* and western quoll *Dasyurus geoffroii* are both known to have been present in the Great Victoria Desert around the turn of the century. They are now restricted to the south-west corner of Western Australia. Maurice recorded the numbat or 'striped anteater' at Ooldea and Tallaringa during two of his desert expeditions.[1,21] A specimen was also collected by Richard Helms in the Everard Ranges on the desert's north-eastern edge during the 1891 Elder Expedition.[22] Aboriginal people recall the presence of both the numbat and western quoll in the central-western Great Victoria Desert during their living memory.[11] The quoll or 'native cat' also featured prominently in important mythological tracks across the desert.[1]

Bandicoots and bilbies have suffered a severe decline in the Great Victoria Desert since European settlement.[23] In the 1920s Bolam commented:

> *...the Ooldea sandhills and the Nullarbor Plain abound in many species of the bandicoot family...a few years ago these valuable little animals were to be seen all over the countryside.*[3]

The story is much different today.

The pig-footed bandicoot *Chaeropus ecaudatus*, so-named because it had only two functional toes on its forefoot and bore a superficial resemblance to the cloven trotter of a pig, was apparently a rare animal within its range even before European colonisation. It is now believed to be extinct across Australia. Maurice commented that the pig-footed bandicoot was common at Ooldea, Waldana and Punthanna and described the animal as 'possessing front feet like the common pig'.[1,21] Bolam never saw a live specimen, but noted that 'its tracks had been seen and followed to its burrow' in the sandhills around Ooldea.[3] According to Aboriginal knowledge, it vanished from the Great Victoria Desert around 60–70 years ago.[11]

The western-barred bandicoot *Perameles bougainville*, desert bandicoot *Perameles eremiana*, golden bandicoot *Isoodon auratus* and greater bilby *Macrotis lagotis* are also believed to be locally extinct in the Great Victoria Desert region.[4]

A dozen specimens of the western-barred bandicoot were obtained by Daisy Bates and Frederick Wood Jones from the Ooldea region between 1921 and 1922.[2] These 'zebra rats', as they were called at the time, were also believed to be plentiful in the Ooldea and Nullarbor areas between 1927 to 1930.[2] The South Australian Museum has three specimens of the western-barred bandicoot, which were all collected in 1931 from Mt Crombie, south of the western Musgrave Ranges and on the northern edge of the Great Victoria Desert.[19] No further museum specimens have been collected anywhere in Australia since this period. The western-barred bandicoot now exists only on islands off the Western Australian coast near Shark Bay.[15] Its common name is derived from alternating light and dark bars on its rump. While occupying a wide range of habitats including wooded sandridges, this bandicoot was probably more common in the saltbush and bluebush plains on the fringe of the Great Victoria Desert rather than in the desert's core.

Two specimens of the desert bandicoot were collected by Richard Maurice for the South Australian Museum around the turn of the century.[1,23] One specimen came from a site between Oolarinna and the Everard Ranges. The site location of the second specimen is unclear, but it may have been obtained on Maurice's expedition to the Tallaringa region in 1903.[1] Finlayson believed the desert bandicoot was common in the north-west of South Australia in the early 1930s.[20]

A single specimen of the golden bandicoot was obtained from Ooldea in 1922, although it is possible the animal could have been brought to Ooldea by Aboriginal people from the far north-west of South Australia. Aboriginal knowledge suggests that the golden bandicoot may have become extinct in the northern fringes of the Great Victoria Desert as recently as the 1950s.[11] Today stable populations exist only on Barrow Island, west of Dampier in Western Australia, and on the Pellew Islands in the Gulf of Carpentaria.[15,17]

The greater bilby or 'rabbit-eared bandicoot' is one of Australia's most delightful mammals. It has silky blue-grey fur, a black and white tail crested throughout its length, long rabbit-like ears and a pointed snout. The bilby is omnivorous by nature and uses its strong forelimbs to dig for food and to construct deep long burrows. Maurice noted that bilbies were common in the Ooldea sandhills around 1900.[1,21] Herbert Basedow, a member of the 1903 Government North-West Expedition (see Chapter 7), recorded the bilby in sandy country near desert oaks skirting the Mann Ranges.[24] Bolam provided several specimens of the bilby for the South Australian

Museum and described it as one of the most common bandicoots in the Ooldea area during the early 1920s.[3] It has not been recorded in the Great Victoria Desert since then, although Jan Turner (Ngaanyatjarra Council) has reported that Aboriginal people captured bilbies from sandhill country south of the Baker Lakes region in 1989 and 1991. (Baker Lakes is just outside the north-western edge of the desert.) A live specimen was also collected from Warburton during March 1977.[5,6] Today its range is restricted mainly to scattered colonies in northern parts of the central and western deserts. Captive breeding programs initiated at the Conservation Commission of the Northern Territory in Alice Springs have proven highly successful. The finding of areas that are both suitable for the reintroduction of captive bilbies and free from predatory cats and foxes and competing rabbits has been much more difficult!

Mortlock Library South Australiana SSL: M: B45287/2

Several specimens of the greater bilby Macrotis lagotis *were collected by Anthony Bolam at Ooldea around the 1920s, one of which is pictured. The bilby is presumed to be extinct over most of the Great Victoria Desert.*

Historical records suggest that the common brushtail possum *Trichosurus vulpecula* existed in the Great Victoria Desert around the turn of the century. While investigating what is now the north-west corner of the Unnamed Conservation Park, several members of the 1891 Elder Expedition came across an elderly Aboriginal woman cooking possum in a fire.[25] The South Australian Museum has a specimen of the brushtail possum which was collected by Richard Maurice in sandhill country between Oolarinna and the Everard Ranges.[1] The species was also apparently common in the range country to the north of the desert, where it lived mainly in red gums lining watercourses.[19]

Many of the desert's early explorers (notably Maurice, Wells, Helms and Carnegie — see Chapter 7) made repeated written reference to animals described variously as 'kangaroo rats', 'spinifex wallabies', 'scrub wallabies', 'native hares' and 'spinifex rats'. They also regularly found 'wallaby fences' or 'wallaby traps' in the desert's sandhill and spinifex country. Aboriginal people used these brushwood structures to capture and then club the 'wallabies' or 'rats' (see Chapter 6). The existence of such enclosures further substantiates the presence of these animals in the Great Victoria Desert and identifies them as an important traditional food source.

Historical descriptions are not detailed enough to enable specific identification of these animals, but it is likely that they were small macropods such as the brush-tailed bettong *Bettongia penicillata*, the burrowing bettong *Bettongia lesueur*, the rufous hare-wallaby or mala *Lagorchestes hirsutus*, and possibly the crescent nail-tail wallaby *Onychogalea lunata*. Maurice collected two brush-tailed bettongs from between Ooldea and Oolarinna, as well as three burrowing bettongs from other parts of the desert and the neighbouring Nullarbor Plain.[1,21] All specimens are housed in the South Australian Museum. The mala was reported to be abundant in the desert's triodia tussocks and numerous in the northern parts of the desert during the 1891 Elder

Expedition.[22,26] The wedge-tailed eagle was recorded preying on the mala during this expedition.[22,26] Basedow recorded mala near Mt Kintore and described them as 'very plentiful in the sandhills' near Mt Crombie on the northern edge of the desert in South Australia.[24] Aboriginal people from the Pitjantjatjara Lands also remember the mala from the region's spinifex country.[19]

On mainland Australia today, the burrowing bettong is presumed extinct, the brush-tailed bettong is found only in three small reserves in south-western Western Australia, while the mala is restricted to isolated colonies in the Tanami Desert.

The lesser stick-nest rat *Leporillus apicalis*, and its close relative the greater stick-nest rat *Leporillus conditor*, built nests consisting of twigs and sticks. They formerly occupied suitable habitats across one-third of the Australian continent.

The chenopod shrubland of the Nullarbor Plain on the southern edge of the Great Victoria Desert appears to have been one of the last strongholds of the greater stick-nest rat. The range of the lesser stick-nest rat is thought to have extended further north across the Great Victoria Desert and into central Australia.[1] The last living specimen of the lesser stick-nest rat was collected by anthropologist Norman Tindale, west of Mt Crombie in 1933.[19] The remains of an old stick nest were found at the base of cliffs fringing the Serpentine Lakes in 1980. It included nest material, droppings and chewed quondong seeds.[7] Old stick nests were found near Point Sunday and Stony Point in the Yeo Lake area during the late 1970s.[6] More than 25 old nests, including two which appeared to contain fresh material, were also found in crevices and caves under breakaway cliffs east of Miss Gibson Hill in the Baker Lakes region. The examination of bone material collected here suggests that these nests were those of the lesser stick-nest rat.[5]

Today the lesser stick-nest rat is presumed extinct. Its larger relative exists as a colony of about 1000 individuals on the Franklin Islands off the South Australian coast.[15] Other colonies have been reintroduced to islands in Shark Bay, Western Australia, and to Reevesby and St Peters Islands in South Australia.[16,17]

Introduced Mammals

Six introduced mammals also occur throughout the Great Victoria Desert. These are the house mouse *Mus domesticus*, European rabbit *Oryctolagus cuniculus*, one-humped or Arabian camel *Camelus dromedarius*, feral cat *Felis catus*, red fox *Vulpes vulpes* and dingo *Canis familiaris dingo*.

The house mouse is ubiquitous across the Great Victoria Desert. Its numbers may irrupt to plague proportions during favourable climatic seasons and following bush-fires.[15] The house mouse was the second most common mammal recorded in the Yellabinna survey, probably as a result of the close proximity of the Yellabinna region to agricultural land.[4]

Rabbits were first introduced into Australia early last century and they began to spread westward and northward across the continent within 50 years. They were first reported on the southern and northern edges of the Great Victoria Desert around the mid-1890s. They had reached Ooldea, Waldana, Boundary Dam and Oolarinna by 1905, and quickly became an important component of desert Aboriginal diet.[1] Today, rabbits are widely distributed and well established throughout the Great

Victoria Desert. During 1980, large numbers were recorded in the desert's heartland around the Serpentine Lakes and near Vokes Hill, where they caused severe damage to vegetation regenerating after fire.[7] Recent breakthroughs in biological control of rabbits offer some hope for a long-term solution to combating this pest. Such controls include a new vector for the transmission of myxomatosis (the Spanish flea *Xenopsylla cunicularis*), the further development of a virus which causes haemorrhagic disease, and genetic engineering studies to inhibit the rabbit's reproductive ability.

The camel was introduced into Australia during the period 1840–1907. It was used widely as the primary mode of transport by explorers of Australia's arid interior and also played a major role as a pack animal in the construction of the Transcontinental Railway and Overland Telegraph lines. Many camels were abandoned following completion of these projects and after the advent of motor vehicles. Feral herds consequently formed and now range freely across the sandy deserts of central and central-western Australia, including the Great Victoria Desert. There is a degree of certainty in stating that the traveller to the Great Victoria Desert will be confronted by the sight of a camel (or herd of camels) bounding down the track in front of his/her vehicle. This initially humorous sight can turn into an exasperating experience when the same camel is still plodding along in front of your vehicle half an hour later! The camel's soft padded feet cause little disturbance to surface topsoil (in contrast to the hard hooves of cattle for example), but its voracious browsing of vegetation and its capacity to drink waterholes dry can cause serious environmental impacts.

There is now growing support for the theory that feral cats were introduced into Australia during the 17th century, when Dutch or Asian vessels ran aground off the Western Australian coast.[27] There is also some evidence to suggest that cats were introduced to the southern Nullarbor coastline near Eyre in the 1830s and 1840s by whalers or sealers.[1] Isolated records of feral cats on the western and northern fringes of the Great Victoria Desert come from the early 1890s.[1] More than 200 cats are known to have been deliberately released on the western Nullarbor in 1899 in an attempt to control the rabbit invasion.[2] Maurice recorded feral cats north and east of Ooldea during the first few years of this century.[1] Today, feral cats range widely across the Great Victoria Desert, although exact numbers are not known.[7] It is now estimated there are around 10 million feral cats in Australia.[27,28]

The survival and persistence of feral cats in arid Australia is due to their independence of free water for drinking, their omnivorous diet, their flexibility in using a variety of habitats for shelter, and their capacity to breed prolifically.[27,28] Biological control, particularly the engineering of viruses which cause self-sterility, offers some future hope for the long-term control of feral cat numbers, although at present this measure appears somewhat of a pipedream.

The red fox is another exotic predator which quickly spread north and west across the Australian mainland after it was introduced into Victoria between 1845 and 1880.[1] It reached the Nullarbor Plain region around 1911–12 and was common in the Ooldea area by the early 1920s.[2] Bolam noted:

> *The fox is causing great havoc among our native fauna...this curse in Australia is of very recent advent to the Trans-Australian wonderland; but it is multiplying at an alarming rate, and the sandhills [around Ooldea] are covered with their tracks.*[3]

Foxes have now been recorded in every conservation park and nature reserve across the Great Victoria Desert. Stomach contents of a fox shot at Neale Junction in the mid-1970s included four sandy inland mice and one spinifex hopping-mouse.[5]

The dingo probably entered Australia at least 4000 years ago, well after Aboriginal people had arrived.[15] It is widespread and common throughout the Great Victoria Desert where rabbits and rodents are its most common prey. A 'dingo-proof fence' now runs from the west coast of South Australia, through large areas of the outback including the south-eastern edge of the Great Victoria Desert, to south-west Queensland in an attempt to stop the dingo from entering sheep grazing lands. It is well known as the longest 'fence' in the world.

Decline of the Desert Native Mammals

The demise of mammals in the Great Victoria Desert (and the Australian arid zone in general) has resulted from the complex interaction of a number of variables, including competition for food by exotic herbivores (notably rabbits), predation by introduced species (such as cats and foxes), climatic factors, and changes in traditional fire regimes.[11,18,29,30]

The rapid spread of rabbits into and across the desert has had a catastrophic effect on native mammals. Rabbits not only compete directly with native species for food and water, but also denude refuge areas that are critical for the survival of native mammals during periods of prolonged drought.

Cats appear to have been present in the Great Victoria Desert (and the arid zone) before European colonisation. These introduced predators may therefore have simply displaced some of the desert's native predators (such as carnivorous dasyurids and owls) or increased the level of predation upon native mammals. While rabbits constitute the main food source for feral cats, native mammal populations may have been decimated in lean times when rabbit numbers were down and alternative prey needed to be found. It may be little coincidence that many of the native mammals that have become extinct are small and within the size range of prey preferred by cats.

Severe droughts were experienced in the Great Victoria Desert and surrounding areas in the late 1890s and in 1914 and 1915. These may have exacerbated the impacts of rabbits and increased predation, further pushing native mammal species to the verge of extinction.

Across the Australian arid zone as a whole, mammals seem to have declined or become extinct from the southern desert areas (including the Great Victoria Desert) first and from the northern areas later. This pattern is consistent with the timing of movement of Aboriginal people from their traditional lands to European settlements.[11] Changes in fire regime, consequent upon the depopulation of desert areas and the ending of traditional patch-burning practices, may have been another reason for the decline of desert mammals.[11,18,29] Patch-burning produced a mosaic of areas of variable age since the last fire, supporting both old and new growth vegetation and providing an environmental diversity which favoured mammals. This scenario may not necessarily be applicable to the entire Great Victoria Desert, as recent studies suggest that fire was used more conservatively in parts of the South Australian section of the desert than in other areas of arid Australia (see Chapters 2,

Mark Shephard

ABOVE: The camel Camelus dromedarius *ranges widely across the Great Victoria Desert and other sandy deserts of Australia.*

RIGHT: The dingo Canis familiaris dingo *will undoubtedly been seen by the visitor to the Great Victoria Desert. It is important not to bury rubbish while camping in the desert, as dingoes will often dig up and scatter your litter.*

Mark Shephard

6 and 12).[31] It is important to note that many Aboriginal people feel directly responsible for the disappearance of particular mammals, believing that it was their abandonment of the desert and the cessation of specific 'increase ceremonies' which resulted in the loss of so many species.[1,11,18] Whatever the reasons for these extinctions, there is a great sense of melancholy when one reflects that these animals will never again grace the desert sands.

Chapter 5

Reptiles

The Desert Reptiles

Regular collections and observations of reptiles were made along the Trans-continental Railway Line during the 1920s and 1930s by prominent scientists such as Frederick Wood Jones, Hedley Finlayson and Norman Tindale, as well as by local identities including Daisy Bates and Anthony Bolam.[1] Their work provided a valuable historic record of the reptiles of the Great Victoria Desert and neighbouring Nullarbor Plain.

Research into the desert's reptile fauna began in earnest during the late 1960s when visiting American herpetologist Eric Pianka conducted a detailed study of lizards in the western Great Victoria Desert.[2,3] His field work, which began in October 1966 and has continued intermittently to as recently as 1990–91, has been significant in revealing the extraordinary diversity of reptiles which the desert

Bynoe's gecko Heteronotia binoei *is widespread throughout arid Australia. It lives beneath logs, loose bark at the base of stumps, and other surface debris.*

Peter Canty

supports. It has also established conclusively that this part of the Great Victoria Desert comprises one of the richest lizard faunas in the world.[4] Remarkably, as many as 47 reptile species have been found living together on a single sandridge site between Laverton and Neale Junction.[4]

Approximately 90 species are now known from the Western Australian section of the Great Victoria Desert, comprising as many as 32 skinks, 16 geckos, 15 dragons, 13 snakes, 5 legless lizards and 5 goannas.[1]

Accumulation of knowledge on the reptile fauna of the eastern Great Victoria Desert has been somewhat slower and it was not until the mid-1980s that the first major compilation of species from this section of the desert was completed.[1] This was based mainly on the register of specimens held by the South Australian Museum, literature review, and field surveys carried out by members of both the Vertebrate Pests Authority (on the southern and eastern margins of the desert from 1978 to 1981) and the Nature Conservation Society of South Australia (from Vokes Hill south to Muckera and west to the Serpentine Lakes during 1980).

Over the past decade, this data has been supplemented by further surveys conducted in the Yellabinna, Tallaringa, Nullarbor, Mabel Creek and Serpentine Lakes regions.[5–10] The Yellabinna wilderness has been the most intensively studied area with 78 species (29 skinks, 14 geckos, 12 dragons, 12 snakes, 7 legless lizards and 4 goannas) known to occur here.[5] A total of around 95 species has now been recorded in the eastern Great Victoria Desert region.[10]

Many of the desert's reptiles have an Australia-wide distribution, for example the sand goanna *Varanus gouldii*, Bynoe's gecko *Heteronotia binoei*, dwarf skink *Menetia greyii* and Burton's legless lizard *Lialis burtonis*. Some occur widely across large areas of arid and semi-arid Australia. These include the dwarf three-toed slider *Lerista muelleri*, smooth knob-tailed gecko *Nephrurus levis*, military dragon *Ctenophorus isolepis* and central netted dragon *Ctenophorus nuchalis*. The Great Victoria Desert comprises a significant proportion of the distribution of species such as the crested dragon *Ctenophorus cristatus*, black-collared dragon *Ctenophorus clayii*, pale knob-tailed gecko *Nephrurus laevissimus* and Centralian coppertail skink *Ctenotus leae*.[1] The linga dragon *Diporiphora linga* and *Diporiphora regina* are endemic to the Great Victoria Desert.[1,5]

Species like the thorny devil *Moloch horridus*, sandplain ctenotus *Ctenotus schomburgkii* and tree dtella *Gehyra variegata* (plus several mentioned above) are ubiquitous throughout the entire Great Victoria Desert. Some tend to replace each other across the desert, from both north to south and east to west. Among the sand dragons, for example, the military dragon is found in the northern desert, but replaced by the mallee dragon *Ctenophorus fordi* in more southern regions. A similar trend is seen with two large yellow-bellied leristas, with the great desert slider *Lerista desertorum* of northern areas being displaced in the south by the myall slider *Lerista edwardsae*.[10] Other lizards may only be found in habitats generally atypical of the Great Victoria Desert. The Peninsula dragon *Ctenophorus fionii* occurs only on the rocky outcrops of Mt Finke in the south-eastern extremity of the desert, while the barking gecko *Nephrurus milii* is limited to areas adjacent to the Nullarbor Plain and Mt Finke.[5]

Although major gaps in the overall knowledge of reptiles still exist, no species occurring in the Great Victoria Desert has a threatened or endangered status when considered on an Australia-wide basis.[1]

Reptile Distribution and Diversity

The wide diversity among the reptiles of the Great Victoria Desert is due to a variety of factors. These include the influence of climate, differences in habitat and soil preference, and variations in lifestyle habits. Together they are responsible for the partitioning of species into separate communities and specific ecological niches within those communities.

As an example of the influence of climate, there is a gradual but definite change in the composition of reptile communities from the more mesic south-eastern part of Yellabinna to the more arid north-west part of the region. The greatest degree of dissimilarity between communities occurs at these two geographic extremes.[5] This change is attributed primarily to a climatic gradient in rainfall which exists across Yellabinna from south-east to north-west. Reptile communities in the south-east live in denser areas of mallee that receive higher rainfall and they do not penetrate the more arid, sparsely vegetated north-western areas of Yellabinna which have less rain.[5] The converse is also true.

Variations in overall physical structure of the environment, as well as changes in vegetation and soil structure, also affect the composition of reptile assemblages.

A number of species exhibit a strong preference for dunefield habitats over other desert landforms. These include the sandhill ctenotus *Ctenotus brooksi*, Centralian coppertail skink, great desert slider, eastern two-toed slider *Lerista labialis*, and pale knob-tailed gecko. Some partitioning of species is also observed between the looser dune crests and more compact interdune areas, with the eastern desert ctenotus *Ctenotus regius*, sandplain ctenotus, fat-tailed gecko *Diplodactylus conspicillatus*, dwarf bearded dragon *Pogona minor*, military dragon and broad-banded sandswimmer *Eremiascincus richardsonii* generally restricted to the swales.[5,10]

In terms of vegetation, spinifex (*Triodia*) grassland provides suitable habitat for a variety of desert reptiles. Many species are adapted to life amongst or beneath the spinifex hummock, which provides refuge, shelter and a gregarious source of food (termites). Examples of reptiles which show a distinct preference for spinifex include the jewelled gecko *Strophurus elderi*, southern spinifex ctenotus *Ctenotus atlas*, leopard skink *Ctenotus pantherinus*, many-lined ctenotus *Ctenotus quattuordecimlineatus*, dusky ctenotus *Ctenotus helenae* and linga dragon.[1,5,10]

Species restricted more to mallee woodlands comprise the western stone gecko *Diplodactylus granariensis*, star knob-tailed gecko *Nephrurus stellatus*, and mallee dragon.[1,10] Those preferring mulga woodland include the smooth knob-tailed gecko *Nephrurus levis*, common desert ctenotus *Ctenotus leonhardii*, and eastern desert ctenotus.

The painted dragon *Ctenophorus pictus* and the saltlake dragon *Ctenophorus salinarum* are two closely related species often found in association with low chenopod or samphire shrublands on the edge of the desert's salt lakes (notably the Wyola and Serpentine Lakes respectively).[10]

Soil texture is another element responsible for partitioning the desert's reptiles. Many of the lizards considered ubiquitous in the eastern Great Victoria Desert can be sub-divided into those which prefer sandy soils (for example the beaded gecko *Diplodactylus damaeum*, thorny devil, sand goanna and rough-nosed blind snake *Ramphotyphlops bituberculatus*), and those which prefer more loamy soils (mulga snake

Peter Canty

Peter Canty

ABOVE: The dwarf bearded dragon Pogona minor *lives mainly in the interdune corridors of the Great Victoria Desert's dunefields. Its 'beard' is small and poorly developed.*

LEFT: The fat-tailed gecko Diplodactylus conspicillatus, *as its name suggests, has a short and plump tail. It occasionally shelters in spider burrows and feeds almost exclusively on termites.*

Pseudechis australis, fat-tailed gecko and sandplain gecko *Diplodactylus stenodactylus*). As mentioned earlier, the common desert ctenotus and eastern desert ctenotus both occur in mulga woodlands. The former prefers mulga on loamy soils, while the latter is found more commonly in mulga among sandy soils. Therefore, even in a continuous habitat like mulga, there can be subtle shifts in the occurrence of species, depending on soil type.[10]

The differing lifestyle habits of reptiles also have an effect on the ecological niche which they occupy. These habits may be categorised into temporal and spatial components. Some reptiles (for example geckos) are nocturnal, while others are more active during the day.[2] Peak foraging time may be restricted to once per day or may be bimodal depending on the season.[3] Foraging technique may also vary. In a spinifex habitat, for example, the desert ctenotus prefers to forage in open space between hummocks, while others like the legless lizards, the southern spinifex ctenotus and the leopard skink restrict their feeding to within or at the edges of a spinifex clump.[3] Most of the desert's reptiles are terrestrial, but a number of species are mainly arboreal, spending much of their day climbing eucalypts and tall vegetation and thereby exploiting niches unavailable to other reptiles. Examples of arboreal species include the tree dtella, pygmy mulga goanna, black-headed pygmy goanna *Varanus tristis* and wall skink *Cryptoblepharus plagiocephalus*.[5]

Peter Canty

The desert pygmy goanna Varanus eremius *is a small, ground-dwelling monitor which is active throughout the year. It covers considerable distances while foraging for large insects (especially grasshoppers) and other lizards.*

Among the surface-dwelling species, the source of ground cover will also influence distribution and diversity. The utilisation of hummock grassland has already been discussed. A number of small reptiles also live among accumulated, decaying leaf litter and fallen timber beneath eucalypts, acacias and other shrubs. Members of the sandswimming *Lerista* genus are generally fossorial (burrowing) by nature and are preyed upon by the desert banded snake *Simoselaps bertholdi*.[5,10]

Desert Reptile Families

There are two major groups of reptiles, namely lizards and snakes. Lizards comprise five main families: geckos, legless lizards, dragons, goannas or monitors, and skinks.[11–14] Snakes may be venomous (belonging to the family Elapidae) or non-venomous (comprising two families, Typhlopidae and Boidae).

Geckos are nocturnal, soft-bodied lizards which have a thin, small-scaled, often translucent skin and large bulging eyes. The scales lie alongside each other and do not overlap, unlike those of other lizards. The pupil of the eye takes the form of a narrow, vertical slit during daylight and expands to fill the eye at night. Geckos lack a moveable eyelid, but have a protective transparent membrane (fixed eyelid) to cover the eye. A broad, fleshy tongue is used to constantly wipe the eye and keep the membrane clean. An interesting characteristic of geckos, as well as legless lizards and skinks, is their ability to discard or 'drop' their tail. This usually occurs in response to attack by a predator or as a result of being alarmed. A replacement tail is soon grown. Geckos are generally insectivorous in their feeding habits, although members of the *Nephrurus* genus will eat small geckos.

The fat-tailed gecko is the most wide ranging of all the genus *Diplodactylus*.[1] It is reddish-brown overall with darker brown markings covering most of its body. It has a short, broad, heavily scaled tail which is used to block access to its burrow where it lives during the daylight hours.[12]

The jewelled gecko is one of the most attractive Australian geckos. It grows to around 10 cm in length and is grey with numerous black-edged cream or pearl-coloured spots along its body. When disturbed, it is capable of exuding a smelly, sticky treacle-like fluid from pores along its tail.[13] Several members of the spiny-tailed geckos (genus *Strophurus*) also exhibit this trait. The jewelled gecko is a spinifex specialist. Its movement among the spiny spinifex thickets is distinctively slow-moving and deliberate.

The arboreal tree dtella is abundant in woodland habitats across the Great Victoria Desert.[15] It is grey overall with irregular black markings, making it ideally camouflaged in and around the loose bark of trees. It has broad pads on its feet which enable it to climb upwards around the trunk of a tree when disturbed.[12,13]

Bynoe's gecko is one of Australia's most adaptable lizards being found across the length and breadth of Australia. Although occurring in a wide variety of habitats, it prefers rocky hills and is less common in sand. It is extremely variable in form, with its body colour often resembling a similar shade to the soil or rocks among which it lives. Interestingly, many of the populations of Bynoe's gecko found in the Great Victoria Desert are composed entirely of females which can reproduce without males.[10]

Knob-tailed geckos (genus *Nephrurus*) are a distinctive, arid-adapted, terrestrial group of lizards which have large heads and short, fat tails terminating in a small knob.[11] Its representatives in the Great Victoria Desert have varying distributions. Some are restricted to southern mallee margins (*Nephrurus stellatus*). Others are centred mainly on the sandy deserts (*N. laevissimus*), or are widespread across the entire arid zone (*N. levis*).

The legless lizards are closely related to geckos and are unique to the Australasian region. All species are long, slender and lack forelimbs, thereby superficially resembling snakes. However, they do possess remnants of hindlimbs concealed in small, scaly flaps of skin which are pressed flat against the sides of their vent and not easily seen. For these reasons, this group of lizards is also commonly known as snake-lizards or 'flapfoot' lizards.

Burton's legless lizard is ubiquitous not only in the Great Victoria Desert but across the whole of Australia. It is highly variable in colour and pattern throughout its range, but those found in the Great Victoria Desert tend to be grey, with or without a lined pattern. It is also diurnal in habit and feeds almost exclusively on small skinks.[11]

The spinifex snake-lizard *Delma butleri* is a triodia specialist and is found throughout the desert where this grass occurs. Its diet comprises a wide variety of insects. The black-headed scaly-foot *Pygopus nigriceps* feeds entirely on arachnids, particularly hole-dwelling scorpions and spiders.[4] It has powerful teeth which can readily disable its dangerous prey.

The dragon lizards (family Agamidae) are active, diurnal, mainly terrestrial lizards whose body is covered with dull, rough scales. In the extreme form, these scales exist as spines. Dragons also have strongly developed limbs and claws, and long tails which can be raised and used as a balance when running. Unlike some reptile groups, dragons do not shed their tails. All are strictly diurnal in their feeding habits. Smaller desert species subsist primarily on ants. Medium-sized dragons eat mainly insects and other arthropods, while the larger central bearded dragon *Pogona vitticeps* is primarily herbivorous.

The slow-moving thorny devil is perhaps the most well-known agamid. Its body is covered by thorny spines and sharp scaly ridges. Despite a somewhat fearsome outward appearance, it is actually harmless. Its jaws are very weak and contain interlocking teeth which are used to slice the 1000 or more ants which it can consume in a single meal.[13] Perhaps the thorny devil's most interesting feature is the large spiny hump projecting from its neck. The function of the hump remains unknown, although it has been suggested it may act either as a store of fatty tissue when food is scarce or as a false head to divert a predator's attention.[13] Another unique characteristic of the devil is its ability to take up water over the surface of its skin by capillary action through a series of microscopic channels between its scales.[13] Like most reptiles in the Great Victoria Desert, mating generally occurs around spring. The female thorny devil lays around eight eggs in a small burrow which is made in sandy soil.[16]

Many dragons display complex social and courtship behaviour, including 'head-bobbing' and 'arm-waving'. The mallee dragon is confined in its distribution to areas of spinifex (*Triodia*) and mallee. It exhibits one of the most unusual social behaviours, which has been described as follows:

> *During the mating season the females react to the presence of a male by raising their hindquarters high off the ground, while the tail is held in a long, high arc. The females orientate their bodies so that the anal region is always facing the male, and if the male moves away then the female will often follow him, always holding her body in the same pose. An interested male will rush up to the female and 'taste' her hindquarters. Whether the female's behaviour constitutes acceptance or rejection of the male is not known with certainty, but as mating rarely follows this behaviour, rejection seems its likely purpose.*[13]

In association with these visual displays, the males of some *Ctenophorus* species (for example the painted and military dragons) are often much more brightly coloured or strongly patterned than the females.

Members of the genus *Ctenophorus* also occupy a variety of different niches in the Great Victoria Desert. The painted dragon is often found near samphire around salt lakes. The Peninsula dragon is restricted to the rocky slopes of Mt Finke. The military dragon and comparatively duller central netted dragon are widespread arid zone species. The crested dragon, which is widespread in the Great Victoria Desert, is also known as the 'bicycle dragon' because of its ability to stride at high speed on its hind legs only, the forelegs being pressed into the body and the tail held semi-erect for balance.[17]

The linga dragon is a long-tailed, buff-brown dragon which has most of its geographic range within the Yellabinna region of the south-eastern Great Victoria Desert.[5] Here it may be found on sandridges dominated by spinifex grass (*Triodia*) and mallee eucalypts. It lives principally in and around spinifex clumps and is most commonly seen perched on top of tussocks.[17] During the Yellabinna survey, it was recorded at Ooldea and in the Yarle and Pinjarra Lakes areas.[5] *Diporiphora regina* is closely related to the linga dragon and replaces the latter on the Western Australian side of the desert.

Goannas or monitors, belonging to the family Varanidae, are fast-moving, powerful lizards with long necks and tails, tough leathery skin, deeply forked tongues and

Peter Canty

ABOVE: The alert, active Burton's legless lizard Lialis burtonis *is Australia's most well-known and widely distributed legless lizard.*

LEFT: The thorny devil Moloch horridus *moves across the desert sands with a distinctive jerky gait, similar to a clockwork toy. Its common name, devil, results from the outwardly fearsome appearance of this inoffensive dragon.*

Mark Shephard

RIGHT: The male painted dragon Ctenophorus pictus *is strongly patterned and among Australia's most colourful lizards.*

BELOW: The crested dragon Ctenophorus cristatus *has the ability to change its colour rapidly if disturbed or aggravated. It feeds mainly on insects, spiders or small lizards.*

Peter Canty

David Pearson

Peter Canty

Peter Canty

ABOVE: The jewelled gecko Strophurus elderi *lives among spinifex (*Triodia*) hummocks and feeds on a variety of arthropods, notably termites.*

LEFT AND BELOW: The leopard skink Ctenotus pantherinus *and dusky ctenotus* Ctenotus helenae *live together among dense tussocks of spinifex (*Triodia*) grass.* Ctenotus helenae *was named in the late 1960s by Eric Pianka, who is an acknowledged pioneer of research on the ecology of lizards on the WA side of the Great Victoria Desert.*

Peter Canty

blade-like teeth. All goannas are predatory by nature, taking mainly vertebrate prey. Five species have been recorded in the Great Victoria Desert.[18]

The perentie *Varanus giganteus* is the largest Australian lizard, growing in excess of 2 metres. It has been recorded in rocky breakaway and sand dune country on the extreme western and north-western margins of the desert.[18,19]

The sand goanna, also known as Gould's goanna, grows to a length of around 1.3 metres. It is the most common and widespread member of its genus, occurring throughout Australia. It prefers a terrestrial existence and seldom climbs trees. When disturbed, the sand goanna displays an amazing burst of speed and usually disappears down its own burrow or a disused rabbit warren. It feeds voraciously on a variety of prey, ranging from mice, rabbits and birds to small lizards. More than 20 species of lizards have been identified from the stomach contents of sand goannas collected in the Great Victoria Desert.[19]

Three smaller 'pygmy' goannas are known to varying degrees in the Great Victoria Desert. The terrestrial desert pygmy goanna *Varanus eremius* was recorded as common in the western half of the Great Victoria Desert during the late 1960s.[18,20] It is much less well known in South Australia, but recent records suggest it is widespread.[10] This shy, wary goanna does not burrow but may often investigate the fresh diggings of other animals and occasionally use them as a retreat.[17,20]

The pygmy mulga goanna is a widespread arboreal species, which is found on tree trunks under loose bark or in hollows in timbered woodlands. It is common on the eastern edge of the Great Victoria Desert, extending to the Lake Eyre Basin.

The black-headed pygmy goanna is widespread throughout central and northern Australia, but seldom seen in South Australia.[5] On the Western Australian side of the desert, this mainly arboreal goanna was found to be relatively common and most active during October.[21] Other lizards are its major prey and constitute more than 70 per cent of its diet by weight.[18]

Skinks (family Scinidae) are the most widely represented group of lizards in the Great Victoria Desert (around 30 species). The diversity of form, habits and reproduction among skinks demonstrates an evolutionary adaptability unrivalled by other Australian reptiles.

Almost half of the skinks found in the Great Victoria Desert belong to a single genus, *Ctenotus*. The comparative ecology of this group was studied in the western sector of the Great Victoria Desert and a number of dimensions to their niche were identified.[3] Species were partitioned according to differences in habitat preference, body size, foraging techniques, prey size, and time of maximal activity. Up to seven species of *Ctenotus* were shown to coexist together: *Ctenotus calurus, C. grandis, C. helenae, C. pantherinus, C. piankai, C. quattuordecimlineatus* and *C. schomburgkii*.[3]

The sandplain ctenotus is the most widespread and versatile of the desert's skinks and is diurnal by nature. The sandhill ctenotus, common desert ctenotus and leopard skink also have a wide distribution across the Great Victoria Desert and beyond. The sandhill ctenotus was the most numerous skink encountered at the Serpentine Lakes.[1] The blue-tailed skink *Ctenotus calurus*, narrow-lined ctenotus *C. dux*, dusky ctenotus *C. helenae*, as well as *C. colletti*, *C. grandis* and *C. piankai*, are more well known from the western sector of the Great Victoria Desert, although the blue-tailed skink was recently recorded in the Mann Ranges to the north of the desert and the narrow-lined ctenotus has now been collected from the Serpentine Lakes and the Mann Ranges.[10] The manylined ctenotus displaces the more southerly distributed southern spinifex ctenotus in the northern area of the Great Victoria Desert across both States.[1]

Skinks belonging to the genus *Lerista* are particularly unusual because, in many species, the limbs have undergone complete or partial degeneration. These burrowing skinks are often termed 'sandswimmers' or 'sliders' because of their ability to 'slip' through loose sands. The eastern two-toed slider is widespread and abundant throughout the desert's dunefields. It coexists with another common species, the western two-toed slider *Lerista bipes,* in black oak woodland at the Serpentine Lakes.[1]

The dwarf skink is widely distributed throughout most of Australia, with the exception of its northern, eastern and south-eastern extremities. This tiny grey-brown to bronze-brown skink is ubiquitous throughout the Great Victoria Desert where it scurries among leaf litter and grasses.

Snakes found in the Great Victoria Desert belong almost exclusively to two main families, Typhlopidae (the blind snakes) and Elapidae (venomous snakes). The only known example of the Boidae family in the Great Victoria Desert is the woma *Aspidites ramsayi*, which has been recorded near the Wyola Lakes.[10]

Blind snakes are small, burrowing, worm-like snakes that are completely harmless to people. They are nocturnal by nature and feed primarily on ants. Their common name 'blind' refers to their eyes, which are rudimentary.[22] The three main blind snakes found in the Great Victoria Desert are the southern blind snake *Ramphotyphlops australis,* the rough-nosed blind snake and the Centralian blind snake *Ramphotyphlops endoterus*. The first two are widespread across most of South Australia, extending across the desert into the south-western corner of Western Australia. The latter is restricted to the central desert region, including the Great Victoria Desert.

Elapid snakes all possess a pair of hollow fangs at the front of their upper jaw through which venom can be injected into their prey. This group of snakes includes some of the most venomous snakes in the world, although many of the smaller elapids are unlikely to harm a human because they have short fangs, deliver small amounts of venom, or are reluctant to bite when handled.[22] Around 10 species of the elapid family occur in the Great Victoria Desert.

The mulga or king brown snake *Pseudechis australis* and the western brown *Pseudonaja nuchalis* are both highly dangerous species which are widespread across the desert and, indeed, much of Australia.[23] The latter is extremely variable in colour and may grow to around 1.5–2.0 metres in length.[22]

The five-ringed snake *Pseudonaja modesta* is a common desert snake. It is medium sized, is active during the day, and feeds mainly on lizards.

At least five species belonging to the genus *Simoselaps* are found in the Great Victoria Desert. All are small, nocturnal, burrowing snakes which have shovel-like snouts and a distinctive pair of black bars on their head.[11] Although venomous, they generally do not attempt to bite and are regarded as harmless to humans. The half-girdled snake *Simoselaps fasciolatus* has blade-like teeth which are used to great effect in slicing open lizard eggs, its primary food source. The western black-naped snake *Simoselaps bimaculatus* has been observed in only two localities in South Australia (Ooldea and Kingoonya). The closest record in Western Australia is more than

Peter Canty

Peter Canty

ABOVE: The fast-moving, alert western brown snake Pseudonaja nuchalis *is one of Australia's most venomous and abundant snakes. Its venom contains neurotoxic and haemolytic activity.*

LEFT: The half-girdled snake Simoselaps fasciolatus *is a brightly coloured, nocturnal, burrowing snake which may be found in sandy areas of the Great Victoria Desert.*

200 kilometres from the State border.[5] Its present status in the Great Victoria Desert is unclear.

The hooded snake *Suta monachus* is a relatively new addition to the reptile fauna of South Australia. This bright red-brown snake with a black head and nape was recently collected from the Tallaringa area, Vokes Hill and Serpentine Lakes in the eastern Great Victoria Desert.[1,7] It is nocturnal by nature, but feeds on diurnal lizards.

FOOTNOTE

The taxonomy of South Australian reptiles is currently under review. The nomenclature used in this chapter follows the 'Checklist and Common Names of South Australian Reptiles', prepared by Mark Hutchinson, Curator of Reptiles, South Australian Museum, Adelaide.

Chapter 6

Traditional Aboriginal Life Prior to European Contact

The next six chapters of the book provide a chronological overview of the land-use history of the Great Victoria Desert, and begin with an account of traditional Aboriginal life in the desert prior to European contact. While the text of this chapter is written in the past tense to follow a chronological theme, it should be strongly emphasised that the essential elements of traditional life, described in historical perspective, are still widely practised by the desert's present-day Aboriginal people.

Aboriginal Occupation of the Desert

Aboriginal people have inhabited the Australian continent for around 50,000 years. The sandy deserts of Australia's interior were the last niches to be permanently occupied, owing to their inherent harshness and aridity.[1] Aboriginal association with the Great Victoria Desert is traceable to about 20,000 to 24,000 years ago.[2]

Aboriginal people living in the Great Victoria Desert formed (and continue to be) part of the large 'Western Desert Culture Bloc' which encompassed the sandy deserts and range country of central and western Australia.[2–5] Within the broad Western Desert Culture Bloc, all Aboriginal groups shared a common language and a remarkably unified social structure and lifestyle. They also had closely related religious beliefs and practices centred on the 'tracks' taken by totemic ancestors of the *Tjukurpa* or The Dreaming.[6]

Many attempts have been made to delineate various groupings among the Aboriginal people who lived in and around the Great Victoria Desert. Well-known anthropologist Ronald Berndt made the distinction between *pila*, those people occupying the 'spinifex plains' of the Great Victoria Desert, and *yapu*, those inhabiting the 'stony' hill or range country to the north and west of the desert.[3] Another anthropologist, Norman Tindale, classified the Great Victoria Desert's Aboriginal people into a number of groups, including the Kokatha, Nangatadjara, Ngalea and Pindinii;[7] however, the latter two terms are not recognised by Aboriginal people now living in the south-western part of the desert.[4] Other groups, such as the Pitjantjatjara, Yankunytjatjara and Antakarinya, also belonged to the Western Desert Culture Bloc but inhabited the range country to the north and north-east of the Great Victoria Desert.[7]

All groupings were distinguished by minor dialectical differences within a common language structure, rather than by fixed 'tribal' boundaries.[3] The area occupied by a particular dialectical group was fluid and flexible. Population movements in and around the desert were common throughout Aboriginal prehistory. Further migrations occurred following European contact. An extended drought across southern

Australia from 1914 to 1915 forced the Pitjantjatjara people to move eastwards, displacing the resident Yankunytjatjara and Antakarinya further south.[2,3] Subsequent, often enforced, movements to European settlements and missions (see Chapter 8) also caused fragmentation of some of the desert's Aboriginal groups. Today, particularly in the southern part of the desert, past affiliations with a particular dialect or specific area of land have been enveloped within a major cultural amalgamation, and a redefinition of Aboriginal cultural identity has occurred. Most people living at homelands within the southern Great Victoria Desert (Oak Valley and Tjuntjuntjara) or settlements on its periphery (such as Yalata and Coonana) now collectively call themselves 'southern Pitjantjatjara' (see Chapter 11).[2,4,8–9]

Traditional Life in the Desert

A generalised picture of traditional life in the Great Victoria Desert prior to European contact can be woven from the present-day reflections of older Aboriginal people, as well as from the prolific writings of early explorers, surveyors, prospectors, anthropologists (particularly Ronald and Catherine Berndt and Norman Tindale), and others such as Daisy Bates and Anthony Bolam (both intimately connected with Ooldea).

In essence, life revolved around small family groups employing an extremely mobile hunter-gatherer economy.[2] The desert's climate and environment were simply too harsh and unpredictable to sustain large groups of people for long periods. Instead, occupation was based on seasonal cycles of shortage (in dry periods) and abundance (following rain).[2]

Limited hunting and gathering took place during drought. Specialised water conservation and management techniques enabled people to survive in the short term. These included constructing dams on claypans to trap water, obtaining water from the roots of certain trees, and placing brush or stones over rockholes to reduce evaporation and prevent pollution.[10] As the availability of these supplies subsided, Aboriginal people were forced to retreat, often over many hundreds of kilometres, to a limited number of sites with permanent water. The gathering of numerous groups in the one location provided opportunities to participate in broader cultural activities such as ceremonies and trade.

After rain, family groups would disperse opportunistically into the desert's sandhill country, set up 'camps' at transient water supplies and exploit abundant food reserves.

A network of Aboriginal routes or 'highways' provided access to and across the desert.[11,12] Aboriginal people retained an intimate knowledge of such routes, which invariably followed sources of water and were intimately associated with Dreaming tracks and sites of mythological significance. In utilising these routes, groups were able to maintain and reaffirm their religious and spiritual affiliations with the landscape. Many of the routes extended well beyond the boundaries of the Great Victoria Desert, for example to the range country to the north of the desert, to the southern edge of the Nullarbor Plain and Great Australian Bight, and to areas around the present-day towns of Kalgoorlie and Coober Pedy.[6,8] Some modern-day roads follow these traditional Aboriginal 'highways' quite closely, for example, the main Laverton to Warburton road, the Vokes Hill to Cook road, and the recently constructed 'Great South road' from Wingellina to Tjuntjuntjara (see Chapters 7, 10 and 11, respectively).

The Basic Elements of Traditional Desert Life

Three of the major elements of traditional life in the Great Victoria Desert were the people's spiritual affiliation with the land, their dependence on key sites for water, and their hunter-gatherer subsistence economy. These elements remain of fundamental significance to contemporary desert Aboriginal people.

One Law, One People, One Land

Aboriginal people of the Western Desert Culture Bloc have a different concept of Creation and time to that of Europeans. They believe they were descended from Ancestral beings who emerged from the land during a period of Creation known as the *Tjukurpa* or The Dreaming. Some of the beings appeared in human or animal form, while others were described as both human and animal, or human and plant. This does not imply the beings had physical shapes that were, for example, 'half-human and half-animal', but rather that they were equally human and animal in essence.[13,14]

As these Ancestral beings wandered over the land, they shaped the physical features of the environment such as the sandhills, claypans, saltlakes and water sources. They also engaged in similar activities to that practised in traditional Aboriginal life, that is, they gathered food, made weapons, married, sang songs, performed ceremonies and had disputes. Their actions established the fundamental principles, or 'Law', on how life should be lived. *Tjukurpa* provided a narrative of things that once happened and a charter of things that still happen.[14] Europeans often perceive The Dreaming as a shadowy reflection of real life or a time of imaginary activities. To the desert's Aboriginal people nothing could be further from the truth; *Tjukurpa* is reality itself, from which all aspects of physical, social and spiritual life are derived.[14]

Aboriginal people also have a different concept of time. Europeans think of time in a linear sense, with past events having a defined beginning and end. The Aboriginal view of time is circular, in which history is viewed as the regular recurrence of established patterns.

Thus, as 'one people', the desert's Aboriginal people believe that they are born from the earth: while alive they are the land's inherent and permanent custodians who have a moral and spiritual responsibility to conserve its features, plants and animals, and to perpetuate the knowledge of *Tjukurpa* through ceremony and song; at death they again become part of *Tjukurpa*, part of their own creation.[14,15] The circle of perpetual life thus continues. Aboriginal people believe they not only come from the land, they are the land.[15]

Today these principles form the foundation upon which current issues relating to Aboriginal land rights are based (see Chapter 11).

The Pivotal Role of Water

Water was the major limiting factor for the survival of traditional Aboriginal people living in the Great Victoria Desert. Sources of water in the desert included soaks or native wells, rockholes, claypans and associated dams, water-bearing tree roots and dew.[10] As

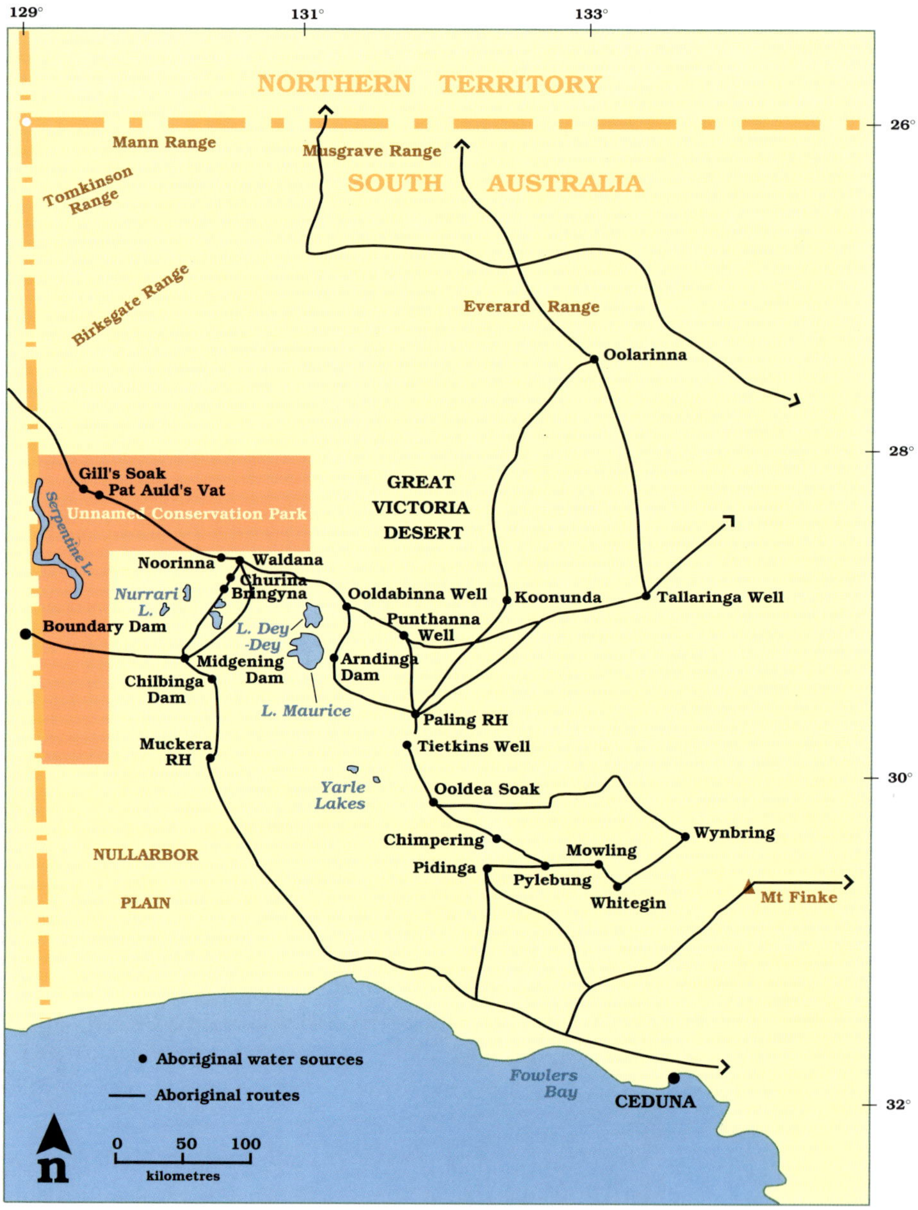

Location of selected Aboriginal water sources (and routes linking these water sources) on the South Australian side of the desert.

many as several hundred locations from which water could be obtained are now known. Sites of permanent water were widely separated nonetheless, and many sources were minor and ephemeral in nature.

Freshwater soaks or native wells are generally located in the centre of low-lying depressions, often between dunes. They consist of coarse-grained sands overlying an impermeable clay layer. Water from direct rainfall or local runoff percolates through the sand until it meets the clay layer where it is then trapped. Aboriginal people obtained water by superficially scraping the surface or by digging narrow shafts or

Mark Shephard

Heinz Moeller

David Pearson

ABOVE: Rockholes of various water holding capacities are scattered throughout the Great Victoria Desert. The rockhole pictured is located in the Yellabinna region of the desert.

ABOVE RIGHT: Aboriginal people used small wooden shovels to construct native dams on claypans throughout the Great Victoria Desert. These dams concentrated the surface water and reduced its evaporation.

RIGHT: A shallow film of muddy water covered the surface of the claypan at Queen Victoria Spring in 1992. An old Aboriginal soak is visible as a sand funnel in the centre foreground of the photograph. This soak was used by numerous European exploration parties.

'wells' to tap the water. The flow of water from a soak or into a well was generally slow, but it was nonetheless continuous and the total water reserve of a given soak was invariably large.[10]

Rockholes (also known as gnamma holes) are usually deep narrow fissures in solid rock which collect local runoff from adjacent surfaces. They are particularly prevalent in the north-western part of the Great Victoria Desert, where they are generally associated with the desert's breakaway country. As mentioned earlier, a rock or brush was often placed over a rockhole by desert Aboriginal people to reduce evaporation by sun and wind and to prevent pollution by animals. The water-holding capacity of rockholes is generally small due to their narrow surface area. As such they were unable to support Aboriginal groups for long periods.[10]

TABLE 7: Examples of different water sources used by traditional Aboriginal people of the Great Victoria Desert

Type	Name	General location	Historical description of water source	Other comments
Soak/well	Waldana (Wandu Well)	Southern desert (40 km south of Vokes Hill Corner)	'At about 7 miles reach Waldana Spring, a little stagnant water showing up in the well. Cleared out ...mud and removed it clear of the well...[its water] is not unpalatable...and has a slight laxative effect. After standing a while a dark sediment forms, and it has a peculiar and unpleasant drying effect on the throat.' (Murray 1904)[27]	Located on a gypseous flat covered with chenopod shrubland. A major water source in the southern Great Victoria Desert, an important meeting place, mythological and ceremonial site. Test drilling by a seismic team in 1965–66 pumped over 8000 litres of clear but mineralised water per day from the well. It was estimated that the well drew from a reservoir at least 3 km^2 in area.[10]
	Tallaringa	Eastern desert (165 km west of Coober Pedy)	'On the 9th day out from Ooldea we reached Tallaringa, a water supply which had never been seen by whites, but which the natives had frequently talked of, although they would not show it to anyone. I found that the surface of the water was fifteen feet below the ground-level, the liquid being contained in a bed of soft, absorbent sandstone. We were able to obtain a thousand gallons a day, a supply which could possibly be very greatly increased by enlarging and slightly deepening the well. This is the most important permanent water known to the blacks for many miles, the nearest permanent water being about two hundred miles away.' (Maurice 1905)[19]	A key water source in the eastern Great Victoria Desert. Functioned as a major cultural focal point and drought refuge centre. The archaeological record of Tallaringa Well and its surrounds has been the subject of two recent studies.[28,29]
Rockholes	Point Sunday	Western desert (south of Lake Yeo, east of Yamarna)	'On the top of Point Sunday there is a large gnamma-hole holding about 400 gallons [1800 litres]. On our outward journey the hole contained 300 gallons [1350 litres], but the natives had put mulga branches in it...to lessen evaporation ...We cleaned the hole out and on our return found about 100 gallons [450 litres] of water, due to a thunderstorm a few days earlier.' (Talbot and Clarke, 1916)[30]	
	Carr-Boyd Rocks	Western desert (south of Lake Yeo, east of Yamarna)	'Some breakaways are seen nine miles to the east, where there are five large rock-holes named Carr Boyd Rock-holes by Hann...After the holes are filled, they will hold water for many months and there is a splendid camp...along the foot of the breakaway.' (Talbot and Clarke, 1916)[30]	
Claypan	Queen Victoria Spring	South-western corner of the desert	'...the water [is] about 150 yards in circumference and from two to three feet deep. Farther up the slopes, at much higher levels, native wells had been sunk in all directions — in each and all of these there was water. One large well, apparently a natural one, lay twelve or thirteen feet higher up than the largest basin, and contained a plentiful supply of pure water.' (Giles 1889)[31]	Located on the western slope of a depression containing a claypan. Ringed by red sand dunes vegetated by tall, open shrubland of mixed acacias, native pine, quandong and native apricot.
Dam	Boundary Dam	Southern desert (Forrest Lakes region in Western Australia)	'...a small native clay-dam on a clay-pan containing a supply of yellow water... The dimensions of this singular little dam were very small: the depth was its most satisfactory feature...the surface [of the dam] was not more than eight feet long, by three feet wide, its shape was elliptical...' (Giles 1889)[31]	

Type	Name	General location	Historical description of water source	Other comments
	Pylebung	Southern desert (90 km south-east of Ooldea)	'...a circular wall or dam of clay, nearly five feet high, with a segment open to the south to admit and retain the rain-water that occasionally flows over the flat into this artificial receptacle...This piece of work was two feet thick on the top of the wall, twenty yards in the length of its sweep, and at the bottom, where the water lodged the embankment was nearly five feet thick. The clay of which this dam was composed had been dug out of the hole in which the water lay, with small native wooden shovels, and piled up to its present dimensions.' (Giles 1889)[31]	
	Paraminna	Southern desert (120 km north-east of Ooldea)	'...on the eastern end of the flat on a clayey patch ... They are of interest as showing the handiwork of the natives. The embankments, shaped somewhat like a horseshoe, are of clay interlaced with mulga twigs, and the blacks must have expended an unusual amount of energy in building them...' (Murray 1904)[27]	
Underground Cavern	Empress Spring	North-western desert (east of Lake Wells and north of Lake Throssell)	'Entrance to cave in surface outcrop of magnesian limestone. Probably permanent soak at end of a subterranean passage leading from cave, bottom of cave 20 ft from surface, soak about 50 ft on underlay ...We watered 9 camels twice, filled water cans and camped here 5 days without lowering the level of the water.' (map accompanying Carnegie, 1896)[32] 'To descend the cave beneath, the natives had made a rough ladder by leaning mulga poles against the edge of the floor. All down the passage to the water little heaps of ashes could be seen where their fires had been placed to light them in their work.' (Carnegie, 1896)[32]	Carnegie states several old camps were scattered throughout the scrub which contained old fires and very primitive shelters formed of a few branches. Animal bones were found amongst the ashes of the fire.

Claypans are low-lying areas which present a relatively impermeable surface to local rainfall. The large surface area of a claypan provides an ideal catchment area for falls of rain, although its shallow depth ensures that evaporation (and minor seepage) rapidly removes any pools of water that form. Claypans were therefore an important, albeit transient, traditional source of water for Aboriginal people of the Great Victoria Desert.

In some instances, Aboriginal people constructed dams on claypans to concentrate the collection of water into a smaller area. The dams were usually low, semicircular walls of clay, and sometimes were interlaced with boughs and foliage to strengthen the dam.[16]

When travelling long distances between sources of water during periods of extended drought, Aboriginal people could still survive in the Great Victoria Desert by collecting dew or by extracting water from the roots of certain trees. The desert's principal water-bearing trees were the red mallee, kurrajong and black oak.[17–20] Aboriginal people simply dug up the lateral roots of these trees, broke them into shorter lengths, and then either stood them on end in a wooden dish to drain or

Peter Canty

ABOVE: Aboriginal people obtained life-saving water from the roots of several trees, including the black oak Casuarina pauper.

LEFT: Ooldea Soak (pictured from the air) was the principle water source for Aboriginal people of the southern Great Victoria Desert, and an important ceremonial and trade centre.

Mark Benbow, Mines and Energy SA

sucked the water directly from the roots.[20] Early explorers often noted piles of black oak or mallee roots at Aboriginal camps, reflecting the importance of these trees both in day-to-day life and as a last resort for desert survival. Richard Maurice (see Chapter 7) reflected on how an Aboriginal companion once saved his life by finding root water:

> *...on one occasion it was entirely owing to [my guide Mungena] that I did not perish miserably in the desert. We had been some days without water, and I, at any rate, was feeling depressed and exhausted. To my surprise Mungena suddenly manifested symptoms of great joy on seeing a peculiar species of tree, known as 'ilburra' [probably the black oak], plainly observable amid the monotonous sea of sand. Taking a little stick he levered up the roots, which he then proceeded to chop into bits. These he drained into our pannikins, and in a short space of time we got four cans full of good water, which saved our lives. The roots of this remarkable tree — invaluable in a thirsty land like Australia — extend from the trunk some twenty or thirty feet in a horizontal direction, and are a little deeper in the ground than those of the mallee tree, from which water is also obtained.*[19]

Empress Spring, in the north-western Great Victoria Desert, is located in an underground limestone cave. The entrance to the cave (pictured) is through a natural sinkhole. Aboriginal people used its waters for thousands of years prior to its European discovery by David Carnegie in 1896. Historical enthusiasts from Kalgoorlie have installed a chain ladder to enable access into the cave.

Mike Donaldson

A detailed account of all the water sources found in the Great Victoria Desert is beyond the scope of this chapter. However, a description of Ooldea Soak, a key and well-documented water source in the southern part of the desert, is provided as an illustrative example of the critical importance of such sites to traditional life. The map on page 90 shows the location of other water sources on the South Australian side of the desert. Table 7 provides brief notes on selected examples of the different types of water sources found throughout the desert.

Ooldea Soak — A Key Water Source

Ooldea Soak is located around 7 kilometres north-north-west of the abandoned railway siding of Ooldea. The soak lies in a circular-shaped depression and contains white sand to a depth of 1–5 metres overlying an impervious clay layer. The depression covers a total area of 5 hectares. It is surrounded by impressive orange-red sand dunes that are well vegetated with mallee, mulga, black oak, sandalwood and a spinifex understorey.[21]

The critical importance of Ooldea Soak to Aboriginal people of the southern Great Victoria Desert and beyond cannot be overemphasised. It has aptly been termed 'an Aboriginal metropolis'.[5] Because of its reliability as a water source, Ooldea Soak was regularly shared by many Aboriginal groups from the southern desert (notably the Kokatha) and

Aboriginal men and women needed to be highly skilled trackers of desert animals. The footprints of a traditional animal food, the Australian bustard Ardeotis australis, *and a more recent addition to Aboriginal diet, the feral cat, are shown on the left and right of the photograph respectively.*

Mark Shephard

neighbouring coastal people, the Wirangu (from Fowlers Bay) and Mirning (from the Eucla area).[21] The latter two groups had some affinity with desert people, but practised significantly different forms of social and local organisation to those of the Western Desert Culture Bloc.[5] Ooldea Soak, known to Aboriginal people as *Yuuldul* or *Juldigabi*, was the focal hub of the network of routes (described earlier) which extended across and outside the desert.[12] All roads literally 'led to Ooldea'. During periods of prolonged drought or when semipermanent supplies in a particular area were exhausted, Aboriginal groups would channel down these routes to seek refuge at the permanent waters of Ooldea. Daisy Bates noted that journeys from the ranges north of the desert to Ooldea often took as long as two years, with people 'zig-zagging in the [Great Victoria] desert for food and water'.[22]

Visiting groups camped in the surrounding dunes in a set fashion, according to the direction of their 'country'. Those arriving from the south occupied sites on the southern side of the soak, while those from the north camped on the northern dunes.[21] Water was available at a variety of locations around Ooldea Soak, but its quality and the depth from which it could be obtained was variable. The northern side of the soak (apparently the site of the 'original' soak) contained the 'freshest, sweetest' water. Another well not far from this area contained 'brackish or bitter' water which was known for its curative powers.[21] Estimates of the volume of water utilised by Aboriginal people are not available. In the early part of this century, it is known that up to 45,000 litres of water per day were pumped from the soak to the nearby Ooldea railway siding (see Chapter 8).

The congregation of people from different regions and cultures ensured that Ooldea became a key ceremonial and trade centre. Richard Maurice, a frequent visitor to the area around the turn of the century, noted that Ooldea Soak was:

> *a permanent native well used by the natives for many generations as a tribal meeting-place. There the blacks would barter, settle disputes, give and take in marriage, and initiate the boys into the mysteries of the tribe. Here in the old days I have seen hundreds of natives assembled at a time.*[23]

Ceremonies were known to attract as many as 400–500 people. They continued to be held at Ooldea until at least the 1940s.[5]

The importance of Ooldea Soak as a trade centre was confirmed by the extent and density of artefact scatter recorded in a recent archaeological survey of the soak.[21] The finding of flaked stone assemblages suggests that Ooldea Soak could have been used by Aboriginal people for up to 6000–7000 years. Historical sources state that items such as spears, boomerangs, shields, native tobacco (genus *Nicotinia*) and wombat fur (used as twine) were commonly traded.[24–26] Quartz and quartzite flakes from the 'hill country' were also exchanged for ochre obtained from a claypan near Pidinga, 50 kilometres south of Ooldea.[12] Other trade items included australites or tektites (known as nulu or 'magic stones' to Aboriginal people), initiation stones, and pearl shells which originated from as far distant as the Pilbara and Kimberley regions.[21]

Ooldea Soak was also of crucial spiritual significance to Aboriginal people. It was an important mythological site and a number of Dreaming tracks passed through Ooldea.[21]

The pivotal role of Ooldea Soak in traditional Aboriginal life began to be undermined as early as the mid-1880s, when initial contact was made with kangaroo shooters and early explorers, and Aboriginal people began to drift towards European settlements along the coast. The poignant story of the Soak's demise following European contact will be described in Chapter 8.

Hunter-Gatherer Subsistence

As mentioned earlier, traditional lifestyle prior to European contact revolved around small family units subsisting on a hunter-gatherer economy. Given the desert's outwardly harsh environment, it could be considered that its original inhabitants were permanently on the verge of starvation. However, as will be outlined, the hunter-gatherer economy had the capacity, particularly in good seasons, to provide a varied, rich and nutritious array of foods.

There was a well-defined division of labour within the family unit. Men hunted for animal or 'meat' foods. Women collected plant or 'vegetable' foods, nectar and other sweet secretions, insects and some small animals.

Most of the animal and plant foods described below continue to be hunted and collected today, particularly by those Aboriginal people now living in the north-western and south-western parts of the desert. Only the tools of the trade have changed, with, for example, rifles and crowbars replacing spears and digging sticks, respectively.

Animal Foods

Traditional animal foods included kangaroos, lizards (especially sand goannas), bettongs, bandicoots, hare-wallabies, emus, bustards, malleefowl, pigeons, parrots, snakes and even dingo pups. Medium-sized to large game, notably kangaroos and hare-wallabies, was hunted by constructing 'traps' or yards.[16] The traps were triangular shaped and built from 'brushwood' (intertwined pieces of shrub and tree branches). Their length and height depended on the size of the game hunted. Larger traps could extend for several hundred metres. Animals were enticed to graze within the confines of the brush fence, before being driven to its apex and then clubbed or speared. This technique was widely used by Aboriginal people across the entire Great Victoria Desert.

Another means of attracting larger game relied on the animal's sense of curiosity. Aboriginal head-dresses comprising brightly coloured cockatoo feathers were bound to the end of a spear or stick and placed vertically in the ground. The animal was drawn to the object and then killed. Emus were often 'drugged' by placing the leaves of pituri (*Duboisia hopwoodii*) into a waterhole.[14] Aboriginal men and women were both highly skilled in tracking goannas and other lizards and digging them up from their underground burrows.

Animal meats (collectively known as *kuka*) were cooked in hot coals and then partitioned according to the status, relationship and age of members of the family group.

Emu and malleefowl eggs were considered a delicacy and widely collected. Eggs and young chicks were also poached from the nests of eagles. To reach the nests, trees were scaled by either slitting notches in their trunks or placing saplings against them.[27]

Rabbits began to infiltrate the desert at the turn of the century and became an increasingly common component of Aboriginal diet. Cats were also hunted, by smoking them out from within tangled brush roots or tree hollows and then clubbing them to death.

George Wood

LEFT: *The ripe red fruit of the quandong* Santalum acuminatum *is a bush delicacy. Its characteristic pitted stone is shown in the top right of the handful of fruits.*

Peter Canty

ABOVE: *Wild tomatoes (genus* Solanum*) are small shrubs with purple flowers, yellow stamens, furry leaves, and fruits like small unripened tomatoes. They spring up after rain and are an important component of desert diet.*

BELOW: *The fruit of the Harlequin mistletoe* Lysiana exocarpi *is sweet, fleshy and eaten raw. Its sticky seed is discarded.*

Peter Canty

Plant Foods

Plants were of fundamental importance to traditional life in the Great Victoria Desert. As well as being a primary food source, plants also provided shelter, firewood and raw materials for domestic and hunting implements. Some plants were used for medicinal and decorative purposes, while others featured strongly in local mythology.[33]

Vegetable plant foods, or *mai*, included seeds, fruits, tubers and roots, and succulent stems and leaves.[18,26,33–35] Seeds provided a staple source of dietary protein. They were collected, winnowed, and ground with water into a paste which could be either eaten raw or baked in hot ashes to produce a damper. Grass seeds were collected by rubbing the seed-heads between the hands. Acacias, such as mulga, witchetty bush, dead finish, umbrella wattle and colony wattle (*Acacia murryana*), formed an important source of seeds collected from trees and shrubs. Desert kurrajong seeds were also widely harvested.

The small black seeds of ruby saltbush *Enchylaena tomentosa* were chewed dry or softened with water. The succulent inland pigweed *Portulaca oleracea* was another plant whose seeds could be gathered and made into small cakes. Entire plants of both pigweed and the closely related parakeelyas were often uprooted and steamed in ashes.

Fruits provided seasonal variety to the diet of the desert's Aboriginals. The fruits of several species of bush tomato (genus *Solanum*) were a popular and readily obtainable food source. Dried fruits of the desert raisin *Solanum centrale* were pounded, moistened with water and made into balls which could be stored and eaten later. The ripe yellow berries of western nightshade *Solanum coactiliferum* contained a bitter juice which had to be extracted prior to consumption. This was achieved by squeezing or piercing the fruit with a stick. The wild gooseberry *Solanum ellipticum* was another edible fruit found in both the desert's sandhills and the range country to the north.

Two members of the genus *Santalum* provided eagerly sought-after fruits. The bright red fruit of the quandong had many uses. Its outer fleshy skin was a rich source of vitamin C and could be eaten raw. If the skin had dried out, it could be reconstituted with water and then consumed. Removal of the skin revealed a pitted stone. A large kernel within the stone was used medicinally to treat aches and pains, and even pounded to prepare a hair conditioner. The native or bush plum produced small blue-black fruits whose flesh could be eaten raw or pulped and mixed with water to provide a thirst-quenching drink.

The small sticky fruits of mistletoe plants, such as *Lysiana murrayi*, *Lysiana exocarpi* and *Amyena maidenii*, were knocked down from trees that were parasitised by the mistletoes, and then eaten raw. They were regarded mainly as 'children's food'.[34]

Other sources of vegetable foods were utilised when food reserves were scarce. Underground bulbs (tubers) of the bush onion *Cyperus bulbosus* were dug from the ground and eaten raw or lightly mashed. The taproots of young kurrajongs could be roasted and eaten. Several species of fungi were also consumed.[33]

The flowers of a number of trees and shrubs yielded a seasonal supply of nectar. This sweet liquid provided a rich source of carbohydrate. Nectar was sucked directly from flowerheads or soaked in water to produce a refreshing 'sugary' drink. The desert grevillea, Latrobe's desert fuschia, corkwood and desert thryptomene represent but a few examples of nectar-bearing plants found in the Great Victoria Desert. Both nectar and dew water were gathered from thryptomene by beating the flowers with wooden collecting dishes (see also Chapter 2). This practice usually occurred early in the morning before sunrise.

Other important food sources resulted from insect activity within certain plants. The witchetty grub is the larvae of the Cossid moth (genus *Xyleutes*). Aboriginal women dug the larvae primarily from the lateral roots of the witchetty bush, but it could also be found among the roots of other acacias (such as the umbrella and colony wattles) and the desert poplar.[14] Witchetty grubs were eaten raw or slightly roasted. They comprised equal parts (by weight) of fat and protein.[36]

Sugary secretions, produced by small sap-sucking scale insects (psyllids), provided an important alternative carbohydrate source for the desert's Aboriginal people.[37] Hard, crusty lerps were found on the leaves of marble gums and other eucalypts of the Great Victoria Desert. Clear sticky honeydew formed as small droplets along mulga trunks and branches. Both secretions were consumed directly, although honeydew was also mixed with water to make a sweet drink.

Insect galls were commonly collected from eucalypt and mulga trees. Larvae burrowing under the bark of these trees cause irritation and swelling around the site, forming a gall or 'apple'.[38] The larvae spend their entire lives inside the gall. Both the larvae and inner gall case were edible, while fluid accumulating in fresh galls could also be drunk. Mulga apples were generally smaller than eucalypt galls and were eaten whole (either raw or cooked in ashes).

Several excellent books detailing traditional desert plants, their Aboriginal names and current uses are now available.[34,35]

Fire as a Management Tool

Throughout Australia's desert regions, fire was (and remains) an integral part of Aboriginal life and culture. Some of its domestic uses included cooking, clearing an area for a campsite, providing warmth and light, and melting spinifex resins to produce adhesives for use with wooden and stone implements. Fire formed part of ceremonies and burial rituals and was a major communication tool between distant groups.[2,17] Patchwork burning or 'fire-stick farming' was used widely for hunting and increasing the productivity/availability of plant and animal foods. The initial clearing of ground as a result of fire enabled animals such as goannas, kangaroos and bustards to be more easily tracked and hunted. Soft nutritious grass shoots, which sprung up shortly after fire, attracted a new regime of animals in the area. Long-term plant regeneration produced a new suite of edible fruits, seeds and tubers.[2]

Aboriginal groups inhabiting the north-western Great Victoria Desert practised patch-burning techniques widely. People took pride in burning their land and considered that they were asserting ownership rights to the land through such burning. Recent research has suggested that former Aboriginal occupants of the Unnamed Conservation Park on the South Australian side of the desert may have burnt the land less frequently and in a more controlled fashion than their north-western counterparts.[2,17] Traditional owners of the Park still living today state that their fathers and grandfathers burnt the land 'only for meat' and attempted at all costs to preserve the desert's vegetation for economic needs.[2,17]

Chapter 7

Seventy-Five Years of European Exploration (1857–1932)

Exploration Begins Around the Desert Edge

European contact with the Great Victoria Desert began in 1857. Stephen Hack and Major Peter Egerton Warburton both crossed the south-eastern extremity of the desert while separately exploring the Gawler Range area.[1–3] Police trooper James Gerharty undertook a 500-kilometre return trip north and north-west from Streaky Bay the following year, during which he also penetrated the south-eastern edge of the desert.[4] On 7 August 1858, John McDouall Stuart camped at and named Mt Finke while en route to Denial Bay. Stuart's initial impression of the desert was not favourable, as shown by his thoughts recorded in his journal such as 'dense scrub as black and dismal as midnight', and 'even worse than Captain Sturt's [Simpson] desert'.[5]

In January 1865 surveyor Alfred De Lisser left Melbourne and headed for the far west coast of South Australia. His dual aims were to sink wells for fresh water near the Fowlers Bay area and to explore the country inland from the Great Australian Bight. De Lisser, and a companion named Hardwicke, left the head of the Bight by horseback on 22 June. They travelled north-west for six days before reaching deep red sands covered with myall, sandalwood and black oak on the southern fringe of the Great Victoria Desert, close to the South Australian–Western Australian border. The pair then retraced their steps back to the Bight, as their horses were exhausted and their water supply was low. De Lisser named the Nullarbor Plain during this journey, stating:

> *The chief feature of the country passed over is the immense plain, which I have called the Nullarbor Plain from its being destitute of any trees.*[6]

In 1868, two well-sinkers, Venning and Howie, became the first Europeans to visit Ooldea Soak. They were led to the soak by a group of Aboriginal people whom they met while working in the Fowlers Bay area. Venning and Howie told the local policeman, Thomas Richards, about Ooldea on their return to the Bay. Richards himself later visited the soak. A blaze 'WH 1868' on a tree near Pidinga, around 50 kilometres south of Ooldea, may have been made by Howie.[7]

In 1874 John Ross undertook an expedition that crossed the eastern margin of the Great Victoria Desert. Ross, who was searching for new water supplies, left the

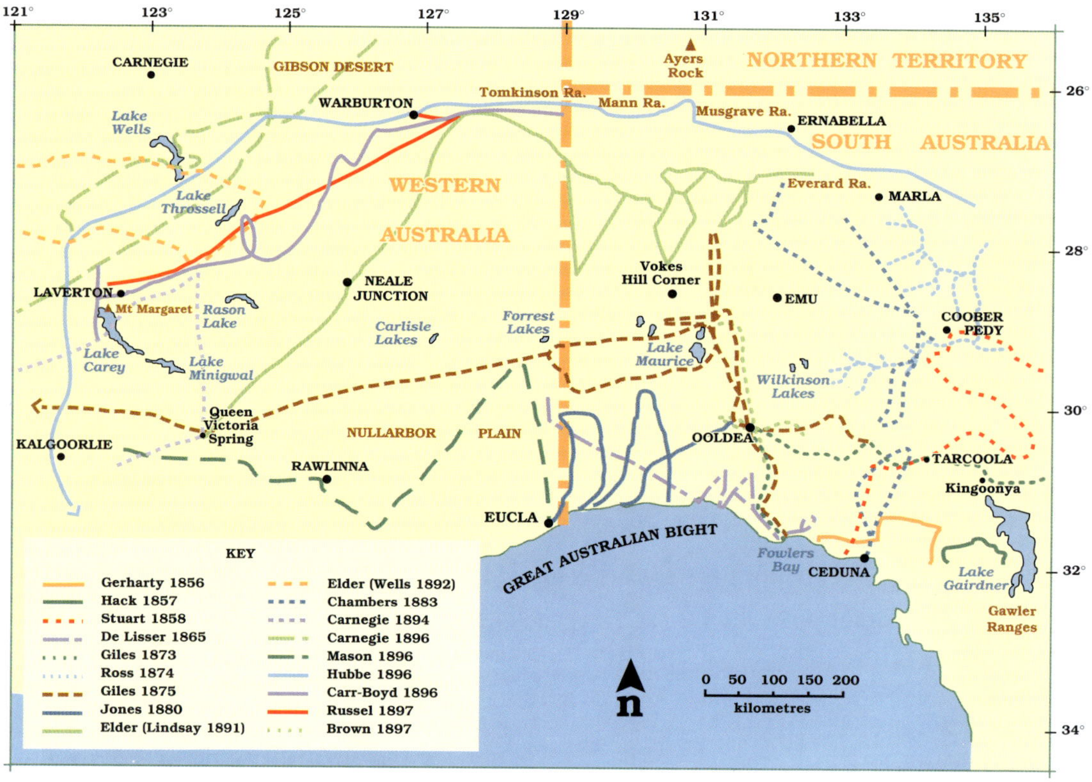

European exploration in the Great Victoria Desert, 1857–1900.

Neales River on March 18 and headed south-west. He arrived at Lake Phillipson almost three months later, and continued to push further west into the desert. Despite making consistent journal references to the 'well-grassed' nature of the country and the presence of native tracks, wurleys, wells and even a dam, Ross was unable to find reliable water.[8]

Ernest Giles and the Romance of Exploration

William Ernest Powell Giles was born in Bristol (England) on 20 July 1835, the eldest son of William and Jane Elizabeth Giles.[9,10] His education at Christ's Hospital in London was interrupted when his parents, brother and five sisters emigrated to South Australia in 1849. He joined his family in Adelaide a year later. After working as a clerk in Melbourne's General Post Office during his early twenties, Giles became restless of suburban life and soon found employment on stations along the upper Darling River. It was during this phase of his life that Giles acquired his skills as a bushman and developed a passion for adventure and discovery. He also befriended William Henry Tietkens, a future companion on two of his expeditions.

Giles set his sights and aspirations firmly on exploring the vast, uncharted block of desert country west of the Overland Telegraph Line, where 'a man could be first to penetrate into this unknown region and where, for a thousand miles in a straight line, no white man's foot had even wandered'.[9]

Mark Shephard

The small township of Fowlers Bay, 135 kilometres west of Ceduna, is bordered by spectacular white sand dunes. Fowlers Bay was the starting point for expeditions into the Great Victoria Desert by Ernest Giles, William Tietkins and Richard Maurice.

Other well-known explorers of the period, namely William Gosse, John Forrest and Peter Egerton Warburton, had set similar goals. The challenge to become the first white man to cross the western part of the continent between the Line and the Western Australian coast began.

Giles undertook a total of five expeditions west of the Line from 1873 to 1876. The narrative of all his journeys, *Australia Twice Traversed*, makes fascinating reading and clearly portrays Giles' romantic 'affair' with exploration.[11] His second expedition (in 1873) skirted in part the range country fringing the northern edge of the Great Victoria Desert, and he and Gosse crossed each other's paths in unsuccessful attempts to push further west. Forrest eventually achieved the first successful crossing in September 1874, travelling from west to east and, en route, also passing along the ranges to the north of the Great Victoria Desert.

Giles' third and fourth expeditions crossed southern sections of the Great Victoria Desert.

The aim of Giles' third expedition was to search for new pastoral country and sites of permanent water to the north and east of Fowlers Bay, before ultimately reaching the cattle station of Beltana in the Flinders Ranges.[12] Beltana was owned by prominent South Australian pastoralist Sir Thomas Elder who sponsored the expedition. The journey was also designed to find a route through this country, by which Giles could later return to commence his fourth and most important expedition along the 29th parallel to Perth.[10,12]

The third expedition left Fowlers Bay on 13 March 1875 and headed north to Ooldea Soak. Giles' party comprised policeman and second-in-command Thomas Richards who led the party to Ooldea, George Murray from Colona Station, cook Peter Nicholls, and several local Aboriginal hands including an older man,

Nanthona, who became better known to the party as 'Jimmy'. Giles had learned of the advantages of camels in desert exploration and equipped the expedition with a pair of dromedaries and a calf, as well as several horses. Upon arriving at Ooldea Soak, Giles remarked:

> *Having reached the northern edge of the plain we had been traversing, we now entered the bed of sand hills and scrub which lay before us...we came in five miles to a spot where...there existed a shallow native well in the sandy ground of a shallow hollow between the red sand hills, and this spot the blacks said was Youldeh.*[11]

Giles, Murray and Jimmy rode to a small rockhole called Paring the next morning. This was the only watering point west of Ooldea with which Jimmy was familiar, but it contained little water and was badly contaminated with rotting animal carcasses and decaying vegetation. Giles made the decision to proceed due east from Ooldea rather than pushing further north. Jimmy knew the country to the east as far as Wynbring, which he had visited as a boy with his mother. He mentioned there were several sources of water en route. His description of Wynbring roughly 'six sleeps from Youldeh [Ooldea]' conjured visions in Giles' mind of a 'large watercourse descending from a mountain'.[11] Giles' enthusiasm for the trek eastward, through what is now the Yellabinna region, was nonetheless dampened by the thought that Jimmy knew nothing of the country past Wynbring. Giles, Jimmy and Peter Nicholls left Ooldea on 24 March with three horses and the camels; the remainder of the party returned to Fowlers Bay. Giles had become somewhat bemused by the contrasting nature of the two adult camels:

> *The bull, Mustara...was a most affectionate creature, and would kiss people all day long; but the Lord help any one who would try to kiss the old cow, for she would cover them all over with — well, we will call it spittle, but it is worse than that.*[11]

The party headed south-east to Chimering rockhole early next morning. The small crevice between two bare, red granite rocks contained only three buckets of 'filthy black fluid' which the horses drank avidly. The party then pushed on a short distance to Pylebung which Jimmy located later that night. Giles was excited on his arrival at the 'large white clay-pan...glistening in the moon's rays, and upon [which], there appeared an astonishing object — something like the wall of an old house or a ruined chimney'.[11] The 'object' was a circular Aboriginal dam one and a half metres in height, the clay for which had been dug with small native wooden shovels.

The country around Pylebung was well grassed and picturesque. Within 400 metres of the dam, Giles found an Aboriginal ceremonial site consisting of a cleared area with a line of stones running between two black oak trees.

The party then continued to a dry Mowling rockhole and onto Whitegin, another rockhole which contained just enough water for the horses. Giles described members of the party as 'choking with thirst and sleepless with anxiety'.[11] The thermometer reached nearly 30°C at dawn the following morning. The expedition was quickly on the move and, after heading north-east, reached Wynbring by mid-morning. Giles' initial impression of the rockhole was in stark contrast to Jimmy's earlier description:

> *...instead of a mountain, it was simply a round bare mass of stone, standing in the centre of an open piece of country...the large creek channel...had microscopicated...to a mere rockhole.*[11]

Giles reflected on the remarkable skills which Jimmy had displayed over the past eight days. He had successfully guided the expedition across 200 kilometres of often thick scrub from Ooldea to Wynbring, yet he had not seen this area since his childhood. The arrival of Giles' party had caused the dispersal of a number of Aboriginal people camped in the area, but Jimmy later brought a small group of five individuals, comprising an old man, his three wives and a baby, to meet Giles.

Mt Finke was visible to the south-east. Giles determined to visit this conspicuous landmark, despite warnings from the old Aboriginal man that it contained no water. The party left Wynbring on 2 April and reached Mt Finke the next day. It proved to be 'the most desolate heap on the face of the earth'.[11]

Giles left the south-eastern extremity of the Great Victoria Desert and headed east-north-east towards Mt Eba. All the horses died in the extreme temperatures experienced over the ensuing days and the party became dependent on the two adult camels for their survival. After travelling for eight days and over 350 kilometres since leaving Wynbring, the expedition finally stumbled upon a small claypan containing murky, but drinkable water. Both men and camels fell to their knees and drank the life-saving water. After such a perilous expedition, the final stages of their 1100-kilometre journey to Finniss Springs and Beltana were blessed without further misadventure.

Giles' fourth expedition had the ambitious aim of finding a new route to Perth following an approximate line of travel along latitude 29°S. The expedition was again organised by Sir Thomas Elder, but this time equipped only with camels. Members of the party included Giles, William Tietkens who had accompanied Giles as second-in-command on an earlier expedition, Peter Nicholls, Alexander Ross, Jess Young, an Afghan camel-driver named Saleh, a new Aboriginal guide from Fowlers Bay called Tommy Oldham, and Giles' former guide, Jimmy, who was returning to his family at the Bay. The party left Beltana on 6 May 1875, and proceeded south to Port Augusta where provisions were collected. Bidding farewell to the seaside town on 23 May, they travelled westward via Lake Hart to Mt Finke and onto Wynbring, where they remained until 26 June.

They were then led by Jimmy along a more northerly route to Ooldea, which passed a number of new water sources including Taloreh, Edoldeh, Cundyeh, Yanderby, Mobing, Bring, Poothraba and Pondoothy. Giles was particularly impressed with Mobing and considered it superior to Wynbring. Water was obtained at Mobing from a series of native wells sunk at the base of a large, bare rounded rock. The party finally arrived at Ooldea Soak on 5 July. There had been considerable Aboriginal activity in the area since their March visit.

Prior to leaving Ooldea, Giles made a short business trip to Fowlers Bay. He was accompanied by Ross and Jimmy, who was looking forward to being reunited with his family. Tragically, two of Jimmy's children had succumbed to a viral epidemic during his absence. Giles and Ross returned to Ooldea on 25 July.

Meanwhile, Tietkens and Young had investigated the country north of Ooldea and discovered Ooldabinna native well, lying near the 29th parallel some 150 kilometres away. On 27 July, Giles moved the entire camp from Ooldea to Ooldabinna, a journey which took five and a half days. Ooldabinna had several wells and was situated in a flat, open, well-grassed area. They rested for three days. Giles then decided the party should split up. Tietkens and Young were sent further north, in search of water and a possible route linking Ooldabinna and the Musgrave Ranges. Giles, Ross and Saleh headed west.

Tietkens and Young found numerous native tracks and encountered a large Aboriginal party, but were unsuccessful in locating a reliable water source. The weary duo decided to return to Ooldabinna, having traversed 150 kilometres of high sandhills and thickly timbered woodland.

Giles' group had simultaneously skirted the northern side of Lake Dey-Dey and the Nurrari Lakes without finding a trace of drinkable water. Saleh was sent back to Ooldabinna after four days' travel through the seemingly endless saltlake country. Giles and Ross continued their lonely trek westward, through country which soon changed to thick unbroken scrub comprising mallee, mulga, black oak, sandalwood and spinifex. Giles remarked:

> *The silence and the solitude of this mighty waste were appalling to the mind, and I almost regretted that I had sworn to conquer it.*[11]

Mark Shephard

In 1875 Ernest Giles headed for Mt Finke in the south-eastern desert against the advice of local Aboriginal people at Wynbring. He found no trace of water at this prominent landmark.

After seven days' travel across 250 kilometres of waterless desert, the pair came to a plot of burnt spinifex and numerous native footprints. They also found a small Aboriginal dam on a claypan. The dam, which was more than 2.4 metres long, 1 metre wide and elliptical in shape, contained sufficient water to quench the thirst of the camels. By Giles' reckoning, the longitude of this location was close to that of the South Australian–Western Australian border. He later named the claypan Boundary Dam.

Giles decided to push further west through more saltlakes for around 65 kilometres, but returned to the dam after being unable to find water. He then decided to return to Ooldabinna, initially by a more south-easterly route, with the dual aim of following the northern edge of the Nullarbor Plain and locating new water supplies.

The party left Boundary Dam on 16 August and reached the edge of the plain after traversing around 80 kilometres. He described the plain as:

> *...beautifully grassed [with] cotton and saltbush...A bicycle could be ridden, I believe, over the whole extent of this plain, which must be 500 or 600 miles long by nearly 200 miles broad.*[11]

Giles then changed his course to the north-east. He eventually rendezvoused with the rest of the party at Ooldabinna on 23 August. Overnight showers enabled water supplies to be replenished and Giles' camels to be watered.

The entire party left Ooldabinna and followed Giles' original westward track to Boundary Dam, which was reached on 3 September. Recent heavy rain had caused the dam to overflow. There was 'now six feet of water in it' which was more than sufficient for the party's needs.

The expedition remained at Boundary Dam for a week, during which time Giles pondered the next phase of the expedition. He decided to head directly for Mt Churchman, the nearest charted landmark some 950 kilometres away. There was no known water in between. Giles offered each member of his party the opportunity to return to Fowlers Bay or Eucla, but 'one and all declared that they would live or die with [him]' in his pursuit across 'probably the worst desert on the face of the earth'.[11] The party headed south-west into the unknown on 10 September. The terrain changed over the ensuing two weeks, from grassy open plains dotted with succulent shrubs to thick scrub comprising marble (or desert) gum, mallee, black oak, cypress pine, desert poplar, quandong, spinifex and grass trees. Giles described the marble gum as 'a gigantic species of mallee, but as it grows singly, it resembles a gum-tree'.[11] The party shot a bustard, collected malleefowl eggs, and divided most of their remaining 360 litres of water among the expedition's 18 camels and one calf. The camp's mood on 25 September was tense. Giles sensed he was being blamed for their predicament.

Tietkens led the camel train the following morning. He sent Tommy to investigate the view from a tall sandhill to the south. Tommy returned triumphantly, shouting he had found water after following the footprints of an emu. The party was led to:

> *a miniature lake lying in the sand, with plenty of that inestimable fluid...not beheld for more than 300 miles.*[11]

Giles proclaimed:

> *In the desert a fountain is springing...The water is no doubt permanent, for it is supplied by the drainage of the sandhills that surround it, and it rests on a substratum of impervious clay.*[11]

He named both the claypan and the surrounding desert after Queen Victoria. The discovery of Queen Victoria Spring proved the expedition's salvation.

A week was enjoyed at the Spring. Young planted seeds of different vegetables, plants and trees, including a Tasmanian blue gum, wattles and a large bamboo. Together with Giles, he collected several species of plants, insects and reptiles. The pair also recorded the area's bird life, which comprised pink cockatoos, magpies, hawks, crows and corellas. Numerous Aboriginal artefacts were found and Aboriginal people were sighted in the vicinity of the Spring.

The fully recuperated party left Queen Victoria Spring and the south-western corner of the Great Victoria Desert on 6 October 1875. They arrived at Mt Churchman 21 days later, having located Ularring native well around 130 kilometres north-west of Kalgoorlie. An outlying sheep station near New Norcia was soon reached. The expedition then triumphantly followed the path of European settlement to Perth in mid-November.

Giles undertook one final expedition the following year, which took him from Geraldton north east to the Ashburton River and then east to the Gibson Desert. He then travelled through the Rawlinson, Musgrave and Everard Ranges on his way back to the Overland Telegraph Line. He later received an Italian knighthood and a gold medal from the Royal Geographical Society for his exploratory feats across

Australia.[12] His last years were spent in the Coolgardie goldfields where he worked in relative obscurity as a clerk. Giles died of bronchial pneumonia on 20 November 1897, aged 62 years.

William Tietkens

William Henry Tietkens was born in London on 30 August 1844. Like Giles, he was educated at Christ's Hospital. Tietkens migrated to Australia in September 1860 and initially lived in Adelaide. He soon moved to Victoria, where he held odd jobs in the Castlemaine goldfields and later obtained more permanent work as a railway clerk in Melbourne. He met Giles in 1864 and, as already described, was second-in-command on Giles' second and fourth expeditions.

While working as a surveyor near Richmond in New South Wales during 1878, Tietkens was given an opportunity to return to the Ooldea region. British businessman, Louis Leisler, whom Tietkens had met a year earlier while staying in England, asked him to supervise the sinking of several wells in the area with a view to opening up the land for pastoral development.

Tietkens left Adelaide by steamship on 26 February 1879 and arrived at Fowlers Bay on 10 May. He headed back to Ooldea Soak on 16 June, equipped with camels, well-sinking equipment and provisions.

Tietkens left the soak with a well-sinker named Palmer and an Aboriginal companion, Joe, on 20 June. Their aim was to investigate the country north of the soak for a potential well site. After much searching, Tietkens selected a site for the well on an open limestone plain some 60 kilometres north of Ooldea. This location later became known as Tietkens Well. Tietkens returned to Ooldea and moved the rest of the party to the new well site.

He then explored the country north of Tietkens Well. After travelling around 65 kilometres, he discovered a small group of hills and a native well. Tietkens named the outcrop 'Leisler Hills' after his benefactor. The native well, known as Punthi (or Punthanna), was revisited by Tietkens in mid-September and described as 'having little water...but the flowers and feed were most luxuriant'.[13] While at Punthi, Tietkens was confronted by a group of 20 Aboriginal people, who led him north-west the following day to a native dam called Winderabbi. The dam contained 'an abundance of water' and Tietkens 'put in a picket with his name, the date and the name of this water'.[13]

Well-sinking operations at Tietkens Well continued to progress slowly. Around the middle of August loose dry sand was encountered at a depth around 12 metres and Tietkens was forced to sack the well-sinkers due to their discontent. Tietkens investigated the country west of the well over the ensuing weeks and 'trac[ed] the Ooldea Range to its termination'.[13] He returned to Fowlers Bay in late September where the services of two new well-sinkers, Nelson and McKay, were enlisted. Work started at a new site at Tietkens Well in late October. After a month and a half of solid work, the well was 24 metres deep but no water had been found. Christmas Day was spent at Ooldea, where Tietkens shared a 'huge damper...a bucketful of peach jam and two buckets of tea' with the resident Aboriginal people.

On December 26 the well-sinkers finally struck water which was 'by no means fresh, but still good water for sheep and horses and cattle'.[13] Tietkens was elated. The well was timbered to a depth of nearly 40 metres by early 1880. However, Tietkens' dream was shattered shortly afterwards when the well's upper timbers collapsed. Another well was dug but granite was struck at 20 metres. Funds were now exhausted. Tietkens was forced to abandon the enterprise, describing the fruitless endeavours over the past two years as the most bitter part of his life.

Tietkens returned to New South Wales in 1882 and married. During 1887 he was invited to read a paper entitled 'The Nullarbor Plains and the West Boundary of the Province' at a meeting of the South Australian Branch of the Royal Geographical Society of Australasia (RGSA).[14] He later became well known for his role in leading the 1889 Central Australian Exploring and Prospecting Association's Expedition. Tietkens returned to survey work with the New South Wales Lands Department until his retirement in 1909. He died at Lithgow on 19 April 1933.[13]

The Search for Water Continued

In 1880 the South Australian Government sent its Chief Surveyor James Jones to the Nullarbor region to select potential new bore sites.[15] Jones travelled from Adelaide by the schooner *Grace Darling* and arrived in Eucla on 5 April.[16] Three separate trips were made over the next four months in and around the Nullarbor area and the southern fringe of the Great Victoria Desert.

Jones spent much of his first trip searching for the remains of two men, Fairie and Wooley, who disappeared on the northern Nullarbor Plain two years earlier. The pair's abandoned wagon was located but no trace of the men was found. During his travels, Jones reached the timbered country on the edge of the Nullarbor Plain and came within a few kilometres of the southern boundary of the Unnamed Conservation Park. New instructions from the Surveyor General awaited him at Eucla and resulted in a second trip during May to 'extend the examination of the Nullarbor Plain'.[16]

Jones' final survey took him across the Nullarbor to Ooldea, which was reached on 23 July. He then proceeded north to Tietkens Well where the two sites dug by Tietkens's party were examined. He stated:

> *The wells are commenced in hard compact limestone with concentric black markings, which gives place to siliceous and calcareous sandstone clays, iron bands, black sand, and white micaceous clay, with small lumps of mundic. Slightly brackish water was struck in the black sand at 100ft., but on going deeper into the white clay the water became very brackish.*[16]

In 1883 another surveyor, C.J. Chambers, travelled into the eastern section of the Great Victoria Desert north-west of Lake Phillipson, as part of a journey which took him from Arlcoodaby and Wilgena (near Tarcoola) to the Everard Ranges. Chambers sank numerous wells in an attempt to open up a stock route through the region and reported on the pastoral potential of the country.[17,18]

Elder Scientific Exploration Expedition

During 1891–2, the Elder Scientific Exploration Expedition traversed large, previously uncharted sections of the Great Victoria Desert, as part of a 12-month, 6900-kilometre journey across the north-west of South Australia and neighbouring Western Australia.

Mortlock Library of South Australiana SSL: M: B497

Members of the 1891–2 Elder Scientific Exploration Expedition: (from left to right) Leech, Wells, Warren, Lindsay, Streich, Elliot, Helms and Gwynne.

The expedition was financed by Sir Thomas Elder and organised by the South Australian Branch of the RGSA. It was the largest exploring party ever assembled in Australia, comprising six officers, four assistants, four Afghan cameleers and 44 camels.[19] The expedition's ambitious objectives were: (i) to systematically 'fill in the extensive blank spaces on the map of Australia' which were between latitudes 15°S and 30°S and outside the routes of previous explorers such as Giles, Warburton, John Forrest and William Gosse; and (ii) to make an 'exhaustive scientific exploration' of these regions.[20]

Thirty-five-year-old David Lindsay was chosen as leader. He possessed a wealth of experience in surveying and prospecting. He had led a Government expedition to Arnhem Land in 1883 and two years later made a private trip to the Gulf of Carpentaria which included an incursion into the Simpson Desert.[20–22] Frederick William Leech was second-in-command. Lawrence Allan Wells was recruited as surveyor on the recommendation of George Woodroffe Goyder. Wells later became leader of the 1896–7 Calvert Scientific Exploring Expedition and the 1903 Government North-West Expedition. (The former expedition traversed the Great Sandy Desert. The latter prospected throughout the range country in the north-west of South Australia and the adjacent Northern Territory, their route taking them, in part, to the very northern edge of the Great Victoria Desert.) Franz Paul Victor Streich was the party's geologist, mineralogist, ethnologist and meteorologist. German-born Richard Helms was selected as naturalist, botanical collector and entomologist. Dr Frederick John Elliot held the position of medical officer and acted as the expedition's official photographer. The four assistants comprised Robert Bowden (who was in charge of the camels), cook Alfred Warren, Robert Gordon Ramsey and Aubrey Percival Gwynne. Hadji Shah Mohomet led the Afghan contingent. In considering the expedition's eventual fate, it is perhaps significant to record that Charles Winnecke had declined leadership of the party prior to Lindsay's appointment, while Leech was selected as deputy following William Tietkens' withdrawal for family reasons.[20]

Simon Coxon

This kurrajong tree, blazed by Larry Wells on 30 June 1891, was rediscovered by Simon Coxon in 1986.

After travelling by train from Adelaide, the Elder Expedition left the lonely railway siding of Warrina in the north of South Australia on 2 May 1891. They initially headed west to Mount Illbillie, the highest point in the Everard Ranges. Robert Bowden fell gravely ill and died en route to the Everards; the loss of his camel-handling expertise proved a major long-term setback for the group.

The expedition, now accompanied by a local Aboriginal man called Billie, continued westward from the Ranges on 6 June 1891. Five days later, Lindsay, Helms and Billie made a southerly detour into the northern Great Victoria Desert to examine the country near the most northerly point reached by Tietkens and Young on Giles' 1875 expedition. The vegetation along this southerly route comprised black oak, mulga and 'giant mallee' (marble gum). The sandhills became taller, more irregular and closer together as they proceeded further south. A native well and several old camp sites were found. Three days later the group encountered 'a mound right in the centre of a valley, with salt water at the foot'.[23] According to Lindsay, it was 'exactly the same' as the mound springs of the Lake Eyre Basin. The location was named Purndu Saltpans. Contact was re-established with the westward-moving main party on 17 June.

The expedition split up again on 27 June. Wells, Ramsay and an Afghan named Mahmoud headed south on a seven-day 'flying trip' into the Great Victoria Desert. The remainder of the party proceeded west to Coffin Hill. Wells' group crossed the 28th parallel two days later and entered what is now the northern boundary of the Unnamed Conservation Park. Wells noticed 'numbers of parrots (ringnecks), minahs, crows and two eagle hawks'. The country featured well-grassed flats of open mulga, patches of spinifex (*Triodia*), marble gums and belts of pine. Towering red sandridges were crossed the following day. Camp was made on the north side of a low sandridge, after passing a flat of kurrajong and marble gum. Wells cut a diamond-shaped blaze on the east and west sides of a kurrajong tree and inscribed 'E.E.E./L.A.W./30.6.91' in the western-facing cut. (This blazed tree was rediscovered 95 years later by Adelaide surveyor and adventurer Simon Coxon.[24]) Wells' camp, at latitude 28°15'47"S, represented the most southerly point reached during this reconnaissance. A north-westerly trek was taken over the next four days. Wells' group rejoined the expedition at Camp 14 near Coffin Hill on 4 July. Three of Lindsay's camels had suffered poisoning during the interim period. Lindsay stated:

> *[I] am quite at a loss to understand what plant is causing the mischief. Mr Helms has searched diligently, but cannot recognise any poison plants.*[23]

Over the next week, the expedition passed westward through beautiful and spectacular range country. Lindsay called this area the Birksgate Range, after Sir Thomas Elder's residence at Glen Osmond in Adelaide. He also named two of the most prominent hills in the vicinity: Mt Watson (after Professor Archibald Watson from the University of Adelaide who assisted with the expedition's formation) and Mt Sir Thomas (after the expedition's generous benefactor).

Meanwhile, Wells, Ramsey and Mahmoud began a second flying trip south-west into the Great Victoria Desert on 10 July. The country was described by Wells as having 'numerous desert gums, few acacias and currant bushes, few poplars [and] dense triodia'.[23] A kurrajong tree was blazed at the evening's camp (latitude 27°47'43"S). On 13 July the party followed fresh 'native' tracks to the recent camp of at least a dozen Aboriginal people. Green tree limbs were plentiful at the camp and a nearby sandstone outcrop contained a rockhole half full of water. Wells estimated the rockhole's capacity to be around 900 litres.

Another temporary Aboriginal camp, with a fire still burning, was encountered the following day. The party surprised both a young girl and an older woman who was 'cooking opossum'.[23] They touched nothing in the camp and continued on their bearing. The country was 'all of a wretched nature — jumble of sandhills, and very uneven for travelling'.[23] Camp was made on 14 July at latitude 28°22'12"S. Wells blazed a marble gum on the south-east side of a sandridge 'E.E.E./L.A.W./14.7.91'. The party set out in a north-north-west direction the following morning to rejoin Lindsay.

The Elder Expedition's leader, David Lindsay, blazed this gum tree near Skirmish Hill on 21 July 1891.

RGSASA Collection

On 16 July, Wells found a splendid rockhole which was 1.8 metres in diameter, 1 metre deep and full of good water. Shortly after, the party was astonished to see the relatively recent tracks of a bullock which had apparently 'strayed from some very distant run'.[23] A wallaby trap was also encountered the next day. After travelling just under 400 kilometres in 13 days, 'all over miserable sandridges', they crossed the border between South Australia and Western Australia and rendezvoused with the main party 3 kilometres north-east of Skirmish Hill late on 22 July. Wells reported to Lindsay:

> *Seeing so many tracks of natives leads me to believe that there are many more rockholes in these sandridges... I also saw many smokes from their hunting fires... There are apparently many natives in this locality.*[23]

The expedition entered drought-stricken country towards the end of July.[21] A new depot was established at Camp 33, 20 kilometres from Mt Squires in the Barrow Range (east of the present day settlement of Warburton). The party split into three groups, each searching in a different direction for water. Leech and Lindsay, who had ventured north-west and west respectively, found the country dry and desolate. Wells' group located a rockhole some 200 kilometres south which held around 1400 litres of water. After discussing the party's predicament with Wells and Leech, Lindsay made the decision to head south-west across the Great Victoria Desert to Giles' Queen Victoria Spring some 632 kilometres away.

The expedition set out on 30 August, with an outwardly unanimous conviction that an appropriate course was being pursued. After six days' travel, Lindsay remarked the desert landscape had taken on a very pleasing aspect, with groves of marble gums in the valleys, native pines, desert thryptomene and spinifex everywhere. They soon arrived at the rockhole which Wells had located on his southerly reconnaissance. Twelve Aboriginal men and numerous women and children had taken up temporary residence. One of the men was no more than 1.2 metres tall with a long flowing beard, while another was around 1.8 metres and very muscular.

Lindsay and Wells were devastated to find the volume of water in the rockhole had been reduced from an estimated 1400 litres to around 400 litres. Thirty of the camels were each given a bucketful of the dirty, bitter water that remained. The next morning two of the young camels became ill.

Camp 47 was made early on the afternoon of 10 September. Lindsay suggested to Dr Elliot that the party's supply of lime-juice be opened to have with their evening drink. Elliot agreed but decided to mix the concentrated juice with whisky and water in a galvanised canteen. The entire party became ill with zinc poisoning soon after drinking the concoction.[21,25] Lindsay made only fleeting reference to the near-disastrous incident in his journal, noting 'We all felt better, but very weak and queer all day'.[23] The location is recorded on the expedition's official map as 'Limejuice Camp'. Ninety-nine years later, John and Peter Streich, the grandsons of the expedition's geologist, relocated Limejuice Camp as part of a re-enactment of the latter stages of the Elder Expedition. They recorded its position as latitude 28°14'6" and longitude 125°54'35".[25]

Lindsay's party reached the half-way point between Mt Squires and Queen Victoria Spring on 11 September. A blaze was cut in a marble gum and photographed by Dr Elliot.

Lindsay estimated the party was still 200 kilometres from Queen Victoria Spring on 15 September. The camels were in relatively good condition, having fed well on succulent parakeelya. Lindsay was particularly impressed with the surrounding marble gum country, which he said gave the area a park-like appearance. Grass trees (*Xanthorrhoea*) were encountered three days later, an impending sign that the expedition was nearing Queen Victoria Spring. The exceptionally dry nature of the country and the absence of any bird or animal life was beginning to make Lindsay anxious about the possibility of finding water at the spring.

The expedition reached Queen Victoria Spring during the late morning of 23 September 1891, two days short of 16 years since Giles' visit. Lindsay's apprehension was well founded as he found its surface dry. He states:

> *I...came to the funnel-shaped hollow, Queen Victoria Spring, but, alas for us, the magnificent spring, which we had come 400 miles to find, had ceased to exist, for no water was visible.*[23]

RGSASA Collection

ABOVE: The Elder Expedition arrived at Queen Victoria Spring on 23 September 1891. They found the claypan dry and worked until late that night to tap its water.

RIGHT: A concrete post was erected at Queen Victoria Spring in 1990 by John and Peter Streich, the grandsons of Victor Streich. The post's exact position is latitude 30°25'50" and longitude 123°34'2".

John Streich

The men immediately proceeded to sink a well. They baled and shovelled until midnight. Around 225 litres of water was extracted before the well collapsed. Earlier during the evening Dr Elliot served a generous nip of whisky to party members and made a complimentary speech, congratulating Lindsay on having led the expedition safely across the desert for such a considerable distance without water. He also commented on how well the party had gelled together and trusted 'amicable relations' would continue to the expedition's end. Elliot's remarks were somewhat hypocritical in light of the overwhelming tension which had developed within the group.

A small supply of water was obtained the following day, until the soak's blue clay bottom was struck at a depth of around 4.5 metres.[19] The camels received their first drink of water for 24 days. Their 632-kilometre trek from Mt Squires to Queen Victoria Spring was one of the longest waterless forced marches in the history of Australian exploration.[26] A marble gum was blazed on the southern side of the spring with the inscription 'E.E.E/D.L/60/23.9.91', '60' referring to the number of their camp.

The Elder Expedition left Queen Victoria Spring (and the Great Victoria Desert) on 25 September. Fraser Range was reached in early October. Lindsay made a quick trip to the telegraph station at Esperance, where he relayed the party's progress to the South Australian Branch of the RGSA in Adelaide and received further instructions. The expedition then headed north-west via Southern Cross to the upper Murchison River. The quarrelling, ill feeling and deceitfulness, which had plagued the expedition since its beginning, reached a climax on the last day of 1891. Elliot, Helms, Streich and Ramsay resigned without reason, but threatened to lay 'grave and serious charges' before the (RGSA) Council.[21] Lindsay telegraphed the Council from Geraldton and was instructed to return to Adelaide immediately.

Larry Wells was now placed in charge of a reduced party which headed east from Annean Station on 17 February 1892. A depot was established on the edge of settlement several days later. Wells left the depot with three new men and reached the north-western edge of the Great Victoria Desert by mid-March. Sandridges, marble gums, spinifex and mallee now dominated the landscape. The discouraging sight of endless red sand dunes extending to the north and east for at least 50 kilometres lay before them. Wells decided to retreat from the desert along a parallel course. On 20 March the party reached a large saltlake which temporarily blocked their path. An initial attempt to cross the lake at a narrow point met with disaster. The camels sank into the muddy surface of the lake and several hours of back-breaking work was required to extricate them. A more suitable crossing was soon found. Empty water kegs were used as punts to float some of the party's provisions across the lake, which now bears its discoverer's name, Lake Wells. On his return to Annean Station in early April, Wells learned by telegram that Sir Thomas Elder had decided to terminate further exploration.

A full-scale inquiry by the Council of the South Australian Branch of the RGSA in Adelaide later exonerated Lindsay from blame over the expedition's demise, but Lindsay remained scarred forever by the sordid affair. Wells and Warren were the only members of the party to emerge from the inquiry with their reputations unscathed. Doubts were later cast by Helms, William Tietkens and respected bushman Christopher Giles as to whether the expedition had actually ever reached Queen Victoria Spring.[21] However, Lindsay's journal notes agree in every detail with Giles' description of the spring. Wells also backed Lindsay implicitly. All doubt was dispelled when the noted explorer David Carnegie visited the spring in 1894 (see next section) and found the gum which Lindsay had blazed.

Despite failing to complete its overall objective, the scientific contribution of the Elder Expedition was nonetheless impressive in some disciplines.[26] More than 150 new species of insects were collected. Specimens of land and fresh water molluscs, lichens, fungi, birds, reptiles and mammals were also taken, the mammals including several now extinct species. Nineteen new plants were among the 700 species gathered. A comprehensive collection of rocks and minerals was made, and detailed geological information was recorded. Ethnographic notes were compiled on Aboriginal people. The expedition mapped more than 20 million hectares of country previously unknown to Europeans.[26]

David Carnegie

David Wynford Carnegie, youngest son of the Earl of Southesk, was born in London on 23 March 1871. He arrived in Australia with his friend Lord Percy Douglas in 1892 and headed for the Coolgardie goldfields.

Two years later, Carnegie embarked on a gold prospecting expedition in the Hampton Plains area north and north-east of Coolgardie. He left Coolgardie on 24 March 1894 with a companion Gus Luck, three camels, a pony and provisions for three months. After failing to find gold on the Hampton Plains, Carnegie decided to push into the Great Victoria Desert and head for Queen Victoria Spring.[27] He was encouraged by Giles' description of the spring and Victor Streich's later report of quartz near Ponton Creek.[28] The pair arrived at the spring on 18 April. The claypan was dry but water was soon obtained after digging out one of the many Aboriginal wells. Among the 'white [marble] gum trees', Carnegie located the blaze made by David Lindsay during the Elder Expedition's visit 23 years earlier. Carnegie nailed a piece of tin, inscribed with his party's details and the date, immediately under Lindsay's blaze. There was considerable evidence of Aboriginal occupation at the spring, including old camps, spears, waddies, coolamons and a corroboree ground.

Carnegie and Luck left Queen Victoria Spring on 22 April and headed due north to Mt Shenton following a possible tract of auriferous country which had been marked on the Elder Expedition's map. They took only 160 litres of water. The hitherto unexplored country consisted of spinifex plains, steep sandridges, marble gum 'forests', mallee scrub and dense thickets of mulga.[27] Numerous piles of mallee roots bore testimony to former Aboriginal presence. Carnegie named Mt Luck and Mt Douglas on 27 April and reached the Cosmo Newbery area five days later. In early May the party arrived at Mt Shenton, where they found a stone cairn built by Larry Wells and an excellent soak from which their thirsty camels were watered. Three days of exploration in the Mt Shenton area, followed by an intensive search of the country between Mt Grant and Mt Margaret, failed to find a trace of gold. Their luck changed on 30 May when a 'fine big reef' was pegged in the Niagra district.[27] They returned to Coolgardie on 22 June, after a 90-day, 1300-kilometre journey which included a large section of unexplored country in the Great Victoria Desert.

Carnegie made a second prospecting trip from November 1894 to February 1895.[27] He found gold near Lake Darlot, but almost died of typhoid fever on his return to Coolgardie to file his claim. Following a brief family visit to England at the end of 1895, he was back in Australia early the following year and eagerly preparing another expedition. His aims were to search for gold-bearing deposits between Coolgardie and the Kimberley, and to find a direct stock route between the southern goldfields and the northern Kimberley cattle stations.

A team for the new expedition was soon assembled; it comprised Carnegie, Godfrey Massie, Joseph Breaden and his Aboriginal companion Warri, and nine camels. The party left Coolgardie on 9 July 1896. They initially headed north via Menzies and Doyle Well, before turning north-east towards the Great Victoria Desert. Marble gums, spinifex and high sandridges were encountered on 31 July, indicating the edge of the desert had been reached. They continued north-east across the Great Victoria Desert for two weeks, without finding a drop of water. Two malleefowl nests were discovered, but they contained no eggs. The weary party eventually stumbled across a lone Aboriginal man, who reluctantly led them to an underground cavern which harboured a plentiful supply of water in a side passage. The party laboured until sunset, baling water for the camels who each drank 75 litres of the precious fluid. Carnegie stated:

I have honoured this hidden well of water by the name of "The Empress Spring". A more appropriate name it could not have, for is it not in the Great Victoria Desert? and was it not in that region that another party was saved by the happy findings of Queen Victoria Spring?[27]

The party remained at Empress Spring from 11 to 15 August before leaving the Great Victoria Desert and moving on to Mt Worsnop and Alexander Spring. They crossed the Gibson and Great Sandy Deserts over the ensuing months, and reached Halls Creek in the southern Kimberley by early December. Their return journey commenced on 22 March 1897. The exhausted party was back in Coolgardie early in August. Carnegie was unsuccessful in finding a stock route or mineral district but, in his own words, he did not 'regret one penny of the cost or one minute of the troubles and labours entailed by it'.[27]

David Carnegie returned to England in 1898 and was awarded the Gill Medal for his exploratory work. He was appointed Assistant Registrar and Magistrate in northern Nigeria a year later. He died on 27 November 1900 at the youthful age of 29, the victim of a poisoned native arrow.

William Carr-Boyd

William Henry Carr-Boyd, explorer, prospector and raconteur, was born at sea in 1852 during his parent's voyage from England to Hobart. He spent considerable time exploring both Central Australia and the Cambridge Gulf in the Kimberley region during the 1870s and 1880s. He was also among the early gold diggers at Coolgardie.[29]

In 1895 Carr-Boyd led a prospecting expedition from Kalgoorlie, north-east across the Great Victoria Desert, and then via the Warburton and Musgrave Ranges to the railway siding of Warrina in northern South Australia.[30]

He left the Kalgoorlie goldfields on 27 May with a companion, Mr A. Woodhouse, three camels and provisions for five months.[30] The men proceeded to Lake Carey near the Mt Margaret goldfields and then headed north for 40 kilometres, where numerous gold prospecting teams were encountered. A north-easterly course soon took Carr-Boyd's party to Mt Shenton. After travelling east to the Yeo Lake area, they turned north-north-east across the Great Victoria Desert towards the Warburton Range and Barlee Springs.

Samuel Hubbe's expedition was met near the Officer Creek in the Musgrave Ranges during February 1896 (see next section) and information was exchanged about the country over which each group had travelled.[31]

Carr-Boyd continued to explore and prospect through the Great Victoria Desert region during the early part of this century. Although a very capable bushman, explorer and prospector, Carr-Boyd is best remembered as one of the great camp-fire entertainers of his generation:

[he had an] engaging personality and unfailing optimism...a fine voice, a wide repertoire of songs and an amazing fund of anecdotes.[29]

He died at Dandenong on 11 May 1925, aged 73 years.

Samuel Hubbe

Samuel Grau Hubbe was born on 1 August 1848 in the Adelaide Hills town of Macclesfield. He joined the South Australian Government Survey Department in September 1880. He held positions of Inspector of Vermin Districts and Government Arbitrator over the next five years. In 1895 Hubbe was appointed leader of a party to discover a stock route from Oodnadatta, in South Australia, to Coolgardie in the Western Australian goldfields region.[31]

The so-named Western Australian Stock Route Expedition comprised Hubbe, William Murray as second-in-command and map maker, Thomas Mahar, T. Langman and I. Vines. Hubbe was instructed by the Surveyor General to take meticulous records on the location of potential watering points and suitable feeding areas for stock.

After railing to Oodnadatta, the expedition left the remote outback post on 11 November 1895 and headed for the Everard and Musgrave Ranges. Ernabella was reached on 24 December. Hubbe had also been asked to investigate an extinct mound spring reported by the Elder Expedition four years earlier. On 13 January 1986, Hubbe, Murray, Langman and an Aboriginal guide named Harry headed south from a base camp near the Musgrave Ranges into the northern Great Victoria Desert. They found the mound nine days later. Hubbe and Murray concluded it was not a 'mound spring' at all, but rather the remains of a low-lying gypseous ridge surrounded by country of unusually high level.[31,32] A bore was sunk at the 'mound'. It yielded extremely salty water, totally unsuitable for stock. Hubbe described the desert country as 'nothing but a rolling sea of mulga and other scrubs, growing on red sandhills and intervening flats...[with] not a single point of interest rising out of the sea to relieve the monotony'.[32] The vegetation comprised 'inferior wattle, bastard wattle, stunted quandong and...poor wire grass'.[32]

After returning to their base camp, the party followed the range country and crossed the South Australian–Western Australian border on 9 April. They continued their westward trip through the Cavenagh, Barrow and Warburton Ranges before reaching Alexander Spring.

The expedition crossed the north-western corner of the Great Victoria Desert in mid-May, travelling via the Ernest Giles Range and Lake Wells. There was considerable evidence of Aboriginal occupation in this area, including 'loose sandstone boulders in long parallelograms, circles, squares and diamonds' and 'a very ancient brush yard...for catching kangaroo rats and other small game'.[32] Marble gums were plentiful. Hubbe named the Ulrich Range after his late father and the De la Poer Range in honour of the late Donald John De la Poer Beresford, clerk to the South Australian Parliament. The party left the Great Victoria Desert in late May. They skirted Lake Carey, where piles of freshwater turtle shells were found at old Aboriginal encampments. They then travelled via Menzies to Coolgardie, which was reached on 15 June 1896.

Hubbe left the goldmining town on 27 August and headed south to the Fraser Range, Eucla and Fowlers Bay. He returned to Adelaide by schooner.

Hubbe reported to the Surveyor General that the 'inferior desert country' precluded the finding of a practical route by which stock could be delivered in a marketable condition to the goldfields. He resumed his work with the Survey Department until 1899 when, at the age of 51, he served in the Boer War in South Africa. He died in action, shot through the heart, on 12 September 1900.[32]

Arthur Mason

Also in 1896, the Western Australian Government sent Arthur Gregory Mason to explore the far south-east of the State and to report on the incursion of rabbits into the region.[33] Mason and his party left Kurnalpi (near Kalgoorlie) in June. They travelled south of Queen Victoria Spring, and then south-east towards Eyre. A north-easterly course took them to the Boundary Dam area, close to the former tracks of Giles (1875) and De Lisser (1865). While camped in the area, Mason's camels were unhobbled and driven off by local Aboriginal people. The party was forced to undertake an exhausting eight-day, 300-kilometre walk south across the southern fringe of the Great Victoria Desert and the Nullarbor Plain to Eucla.[33] Mason reported that the land examined was fine pastoral and agricultural country, but had a very poor supply of surface water.

Henry Brown

The South Australian Government geologist Henry Yorke Lyell Brown conducted limited geological exploration in the southern Great Victoria Desert during 1897.[34] He left Fowlers Bay on 23 April and reached Ooldea nine days later. The nature of the country north of Ooldea was then investigated and visits were made to Tietkens Well, Punthanna and Ooldabinna. Brown described the latter as comprising a few rockholes in argillaceous and jaspery sandstone, one of which was 4.5 metres deep. He stated:

> *They lie in a gully or flat coming down from the sandhills, and, owing to the flat having a rocky bottom, are likely to be filled by any good shower of rain.*[34]

The entire region was experiencing a severe drought during Brown's visit and, after reaching a point around 160 kilometres north of Ooldea, he decided to return to its safe waters. He obtained around 70 litres of water from Punthanna during his retreat.

Brown headed north-west from Ooldea on 18 May, before turning south to the edge of the Nullarbor Plain and completing a 160-kilometre circular route at Pidinga.

Country to the north-east of Pidinga was then examined. Rockholes at Lake Bring and Mobing were visited and 180 litres of water were obtained from the latter. Brown's report states that an unsuccessful search was made to locate Wynbring Rocks, although he records coming to an outcrop of granite several acres in extent from which Mt Finke was visible.[34] Could this have been Wynbring? He then headed eastward out of the desert via Mount Eba to Finniss Springs Station.

Hugh Russel

During 1897, Hugh Russel led a gold prospecting expedition from Coolgardie across the north-western Great Victoria Desert to the Barrow, Cavenagh and Warburton Ranges.[35,36] Russel's party comprised H.V. Smith, Frederick Leech and two Aboriginal men, Fred McGill and John Ponton. They were equipped with eight camels and provisions for eight months. After leaving Coolgardie on 3 May 1897, the party travelled via Mt Margaret and Kilpatrick Well to Mt Shenton and nearby Point Virginia. Numerous small wallabies and patches of blackboys (*Xanthorrhoea*) were seen along this stretch of country. A north-easterly course across the Great Victoria Desert was taken on 23 June. Saltlakes, including Lake Throssell, were soon encountered. Bustards were prevalent in the samphire flats near the lakes. The party then travelled over a 160-kilometre expanse of sandridges dotted with occasional marble gum and mulga. Russel wrote:

> *This travelling over sand-ridges is heart-breaking work, one is always on the lookout for a point de vantage whence to look ahead and see what is in store for us, but always some sand-ridge at the great distance blocks the view.*[36]

The sandridge country changed to undulating ironstone gravel plains by 10 July. A large breakaway with three prominent peaks was found on 12 July. After passing a large saltlake, further belts of sandridges were encountered, occasionally interrupted by breakaways. Russel's party eventually reached the Townsend Ridges and then spent the next two months prospecting in the Barrow, Cavenagh and Warburton Ranges. They retraced their route across the Great Victoria Desert and arrived back at Coolgardie on 13 October, their homeward journey taking 34 days. Russell reported the trip 'was barren of results from a prospecting point of view'.[35]

Richard Maurice

Richard Thelwall Maurice was born in Adelaide in 1859, the son of prominent Eyre Peninsula and Mid-North pastoralist Price Maurice.[15,37] Much of Maurice's childhood was spent working on his father's properties. It was during this period that he developed a passion for the Australian bush. Maurice ventured to the Kimberley goldfields in the 1880s, but became stricken with illness and was forced to return to Adelaide. He soon moved to Fowlers Bay where he initially worked for George Murray of Yalata Station and later became a kangaroo shooter on the Nullarbor Plain. Maurice received a substantial inheritance on his father's death in 1894. Thereafter he devoted most of his time and money to exploring the Great Victoria Desert. Between 1897 and 1903, Richard Maurice funded and led at least eight private expeditions into the southern and eastern Great Victoria Desert, an area he referred to as 'the back country'.[37–42]

Adelaide historian Tom Gara has made a remarkably detailed study of Maurice's life and exploratory work in the desert. Gara states:

> *The significant contribution that [Maurice] made to the exploration of this State received little official recognition at the time and has been largely ignored in most studies of the history of the exploration of this State.*[38]

His claim regarding Maurice's lack of recognition is supported by the fact that Maurice is not even mentioned in two of the most comprehensive books available on Australian exploration.[10,12]

Maurice's manuscripts and photographs nonetheless provide a valuable early resource on the flora, fauna and landscape of the southern and eastern Great Victoria Desert in particular. They also give an important insight into the lifestyles and customs of the Aboriginal people who lived in these areas. As Gara states:

> *Maurice was the only white man to have any substantial contact with the Aborigines who inhabited the South Australian portion of the Great Victoria Desert prior to their migration into European*

settlements...and...Maurice was an excellent bushman, an enthusiastic collector of faunal, botanical and ethnographic specimens, and was kind and sympathetic to the Aboriginal people.[15,38]

Around the turn of the century, Richard Maurice made at least eight private trips into the Great Victoria Desert, an area he referred to as 'the back country'.

Most of Maurice's material is now housed in the Mortlock Library of South Australiana or held among the collections of the South Australian Branch of the RGSA. Mainly as a result of Gara's efforts, Maurice may now receive some of the credit he so richly deserves.

Maurice's first trip to the Great Victoria Desert occurred in 1897. He was accompanied by friend, general hand and cook, Bill Voake, and an Aboriginal guide from Eucla named Mingo. Voake also assisted Maurice on some of his later expeditions. The three men left Eucla in July and travelled via Koonalda and Muckera to the Boundary Dam area. Maurice's aim was to investigate the disappearance of the camels and equipment belonging to Arthur Mason and his party. After an unsuccessful search, they were eventually led to Boundary Dam by a middle-aged Aboriginal man named Injida. The tracks of Giles' camels and the remains of his camp were still visible. Maurice's party was soon joined by more local Aboriginal people. Maurice was then taken north to Iningunna water, where Injida said that up to 75 Aboriginal people were camped. However, a shortage of water forced Maurice to retreat from the desert near the State border around latitude 29°S. His party was led to a saltpan called Carloo at the southern extremity of the Serpentine Lakes and on to Muckera Rockhole. They returned to Fowlers Bay via the edge of the Nullarbor Plain and Ooldea.[37]

Maurice was accompanied by a Fowlers Bay Aboriginal man named Mungena on his next trip to the desert late that same year. The pair visited Punthi native well, approximately 120 kilometres north of Ooldea. Mungena proved a reliable companion to Maurice on this and subsequent expeditions.

In 1898, Maurice and Mungena journeyed from Ooldea to Punthi and on to Oolarinna, a permanent well just south of the Everard Ranges. They then completed what Maurice termed 'a sight-seeing tour' of Central Australia, travelling via Oodnadatta, Charlotte Waters and Hermannsburg before returning to Fowlers Bay.

In September 1899, Maurice, accompanied by two other Fowlers Bay Aboriginal men named Tommy and Albiner, became the first white man to see Waldana Well. His route proceeded north past Ooldea, Paling and Ooldabinna before turning west to Waldana. Maurice stayed with the resident Aboriginal people for several days and learned of three rockholes to the north, one of which he believed may have been visited by the Elder Expedition. He returned to Fowlers Bay on 22 October via Batts Tanks, Ooldea and Pidinga.

Some of the exploration routes of Richard Maurice, 1897–1903.

Maurice and Mungena attempted unsuccessfully to find Tallaringa Well on the desert's eastern edge during 1900. They then headed west across the desert via Punthi, Ooldabinna and Lake Dey-Dey to Waldana. The desert had received heavy falls of rain and many of its saltlakes were full and alive with waterfowl. The pair turned north-west and visited Noorina well and Wilbanina rockhole, with the help of an Aboriginal man whom they had met the previous year. Both these water sources are located north of the Nurrari Lakes, just outside the boundary of the Unnamed Conservation Park. Maurice and Mungena continued their exploration north and found wallaby fences, many old camps, and the tracks of Larry Wells which were made during his southerly reconnaissance in July 1891. They then headed south-west to Boundary Dam and returned to Fowlers Bay via Muckera and Nullarbor Station.

Maurice organised a major expedition from Fowlers Bay to the Rawlinson Range during 1901. He was accompanied by Government surveyor William Murray, Bill

Voake, a camel hand and part-Aboriginal man named Lambert, Lambert's wife Helen, and 11 camels. Murray's role was to accurately fix the locations of important water sources. His official report provides a valuable and detailed record of the trip.[42]

Murray joined Maurice and the rest of the party at Pidinga on 16 May. They headed north to Ooldea, Tietkens Well and Paling rockhole and then east towards Paraminna Dams. Rabbits were widespread in the Ooldea area, but numerous bustards were also seen along this route. Piles of broken black oak roots and brush fences for wallaby hunting provided ample evidence of recent Aboriginal occupation of the area. The party arrived at Paraminna on 27 May. Maurice photographed three small horseshoe-shaped dams, which were constructed of clay and interlaced with mulga twigs.

The party took a north-easterly course towards Punthi, Ooldabinna and Waldana the following day. Bird and animal life was noticeably scarce at Ooldabinna, although old wallaby traps were again numerous. Waldana Well was reached on 7 June. Sightings of wildlife had now improved. Murray recorded 'one bandicoot, a small species of laughing jack [red-backed kingfisher] and a few kangaroo rats' just prior to reaching the well. Zebra finches, bronzewing pigeons and dingoes were seen around the well itself. Maurice's hands had become very badly swollen and scarred. He rested at Waldana, while Murray and Lambert headed north in search of the tree blazed by Larry Wells on 30 June 1891.

The pair reached a high sandhill after a day's travel. Murray stated that the sandhill gave:

> *the most commanding view I have had since leaving Fowler's Bay, though the hill itself is not particularly prominent.*[42]

This location was later named Vokes Hill, in honour of Bill Voake.

They then came upon an area that was stricken by prolonged drought and contained much dead mulga. Murray commented:

> *This strip is the worst I have ever seen, and is even worse than Gibson's Desert.*[42]

Wells' latitude was eventually reached; however, their search of the area was hindered by the fact that Murray had inadvertently left Wells' journal at Pidinga. The tracks of seven camels were encountered on a claypan. Wells' party had fewer camels than this and Murray concluded the tracks belonged to a man named Moseley, who had attempted to find one of Wells' rockholes while travelling from Oodnadatta to the Western Australian goldfields some years earlier.[42] Murray and Lambert returned to Waldana just after dark on 14 June.

The entire party then proceeded north-west past Noorina and Wilbania. On the morning of 20 June the expedition was confronted with dense fog which did not lift until 10am. Maurice followed his tracks of the previous year and arrived at a rockhole which he named Pat Auld's Vat. Pat Auld was a friend of Maurice's, a former member of Stuart's expedition across Australia in 1861-2, and an honorary life member of the South Australian Branch of the RGSA.[37] Voake enthusiastically cleaned out the rockhole which was full of debris. It was 4.5 metres deep and 2 metres in diameter at the surface. Murray estimated the total capacity of this rockhole and a smaller hollow in the same outcrop to be over 20,000 litres.[42]

A fine, warm and cloudless morning greeted the party on 22 June. After about eight kilometres' travel in a west-north-west direction, several shallow soaks were encountered in a strip of low-lying sandy country. There was plentiful growth of rushes around the area, which Murray likened to Ooldea. Murray stated the water was fresh, soft and permanent and believed 'this is one of the places [Aboriginal people] keep until prolonged droughts render the rockholes and small native wells useless'.[42] A flock of zebra finches and eight pink cockatoos drank from the soaks during the party's stay. Maurice named the location Gill's Soak, after the South Australian Under-Treasurer and prominent RGSA Secretary, Thomas Gill.[37]

Murray and Lambert left Gill's Soak on the morning of 23 June and headed north on an unsuccessful two-day search for 'the eastern rockhole of the two noted by Mr Wells on his south trip [during the 1891/92 Elder Expedition]'.[42] The entire party left Gill's Soak on 26 June, noting a conical-shaped sandhill 10 chains to the south would prove an excellent landmark during future attempts to find the soak again. (The only white man known to have visited Gill's Soak since is Larry Wells who, in 1932 and at the age of 72, remarkably led a prospecting expedition from Ooldea, north-west into the Great Victoria Desert, and then via the Mann, Rawlinson and Warburton Ranges to Laverton.)[19,43]

Maurice and Murray subsequently found Wells' (1891) rockhole in a sandstone outcrop the following morning. Its position was recorded as being very close to the 28th parallel. Galahs drinking at the rockhole were shot to supplement the evening stew. Murray estimated the rockhole would hold around 3000 litres of water when full. No trace of the bullock tracks seen by Wells could be found.

After leaving the rockhole, the Maurice expedition continued in a north-westerly direction and passed extensive spinifex plains that resembled wheat fields. They eventually crossed the South Australian–Western Australian border, reaching the Rawlinson Range in mid-July. A south-easterly traverse took the party via the Mann, Musgrave and Everard Ranges to Oolarinna.

After three days' rest, the expedition headed south-east towards Punthanna through country which was unknown to Maurice. The sandhills in the region were much smaller than those to the west, making travel more comfortable. Wildlife was conspicuous, with 'rabbits, kangaroo rats, opossums, curlews, minahs, and cockatoos (red crested)' being recorded on 19 August.[42] A young Aboriginal boy crossed the path of the group the following day. He led Maurice to a nearby luxuriant patch of herbage on which the camels fed. Other local Aboriginal people followed behind the boy and told Maurice that Punthanna was dry. Their party was therefore led by an elderly guide to Koonunda native well, a journey which took two days across monotonous and featureless terrain.

Koonunda contained a fine supply of water and was permanent, according to the Aboriginal people. Murray, who along with Maurice was the first white man to see Koonunda, attached great importance to the well as it lay almost mid-way in a straight line between Ooldea and Oolarinna. Six bustards, several ringneck parrots and a flock of 40 swallows were seen at Koonunda. The party left Koonunda and headed south-west via Paraminna to Paling rockhole. A domestic cat and two kittens were killed by Aboriginal people at Paling and eaten 'with evident relish'. A malleefowl was also taken, which 'lent a little variety to [their] customary rabbit stew'.[42] Numerous bustards were again seen near Ooldea and a fresh malleefowl mound, which contained one egg, was found. Rabbits were plentiful, as evidenced by the many shrubs and trees which had been 'ring-barked'.

The weary expedition left Ooldea on 2 September and headed south to the northern edge of the Nullarbor Plain. The area was carpeted with 'acres upon acres of Sturt pea out in bloom — a really magnificent sight'.[42] Fowlers Bay was reached on 7 September.

In 1902 Maurice and Murray undertook another epic journey from Fowlers Bay across Australia to Cambridge Gulf on the northern Kimberley coast. The pair initially travelled to Central Australia via the eastern edge of the Great Victoria Desert and, in the course of this journey, became the first white men to visit Tallaringa Well.

Sufficient water was obtained at Tallaringa to fill their canteens and water the camels. The water was of good quality, although slightly brackish. Maurice estimated the flow of water from the well could provide around 4500 litres per day. This could be further increased if the well was enlarged and deepened. A large post, with 'TALLARINGA' inscribed on one side and 'RTM' on the reverse side, was positioned near the well on 15 May. Bustards, malleefowl, hawks, crows, a 1.8-metre snake, rabbits and dingoes were recorded at or near Tallaringa during their six-day stay.

Maurice's final expedition into the Great Victoria Desert was made during 1903. This journey took him north to Ooldea and Ooldabinna, east to Tallaringa and Lake Phillipson, and south to Wynbring before returning to Fowlers Bay.[37]

After enduring a decade of strenuous activity in the desert, Maurice sought a more relaxed lifestyle in England for several years. He later returned to Sydney and revisited Fowlers Bay in 1909, where he began organising another desert expedition. However, while drinking at the Bay's Globe Hotel, Maurice collapsed and died of a heart attack at the age of 49. His grave site rests on top of a hill at Yalata Head Station, overlooking the spectacular white sand dunes of Fowlers Bay.

Frank Hann

Frank Hugh Hann was born in Wiltshire, England, in the mid-1840s. He came to Australia with his parents in 1851 and purchased Lawn Hill Station in the Gulf of Carpentaria during 1875. A period of prolonged drought and an outbreak of redwater fever forced him to leave the station penniless in 1896. He then headed overland for Western Australia.

Hann initially explored around the Nullagine area of the Pilbara and as far out as Lake Disappointment. He discovered a route into the West Kimberley district via the Leopold Ranges in 1898, and then spent several years trying to raise capital to stock the country he had found. He also named numerous geographical features in the region.

Hann undertook an exploration trip from Menzies to Ravensthorpe and back to Southern Cross in 1901, followed by several trips around the North-Eastern Goldfields area (Menzies, Laverton, Darlot and Wiluna).

Frank Hann's close association with the Great Victoria Desert began in 1903. Between April and August of that year, he made a return journey from Laverton across the north-western part of the desert to the Warburton Range. His outward

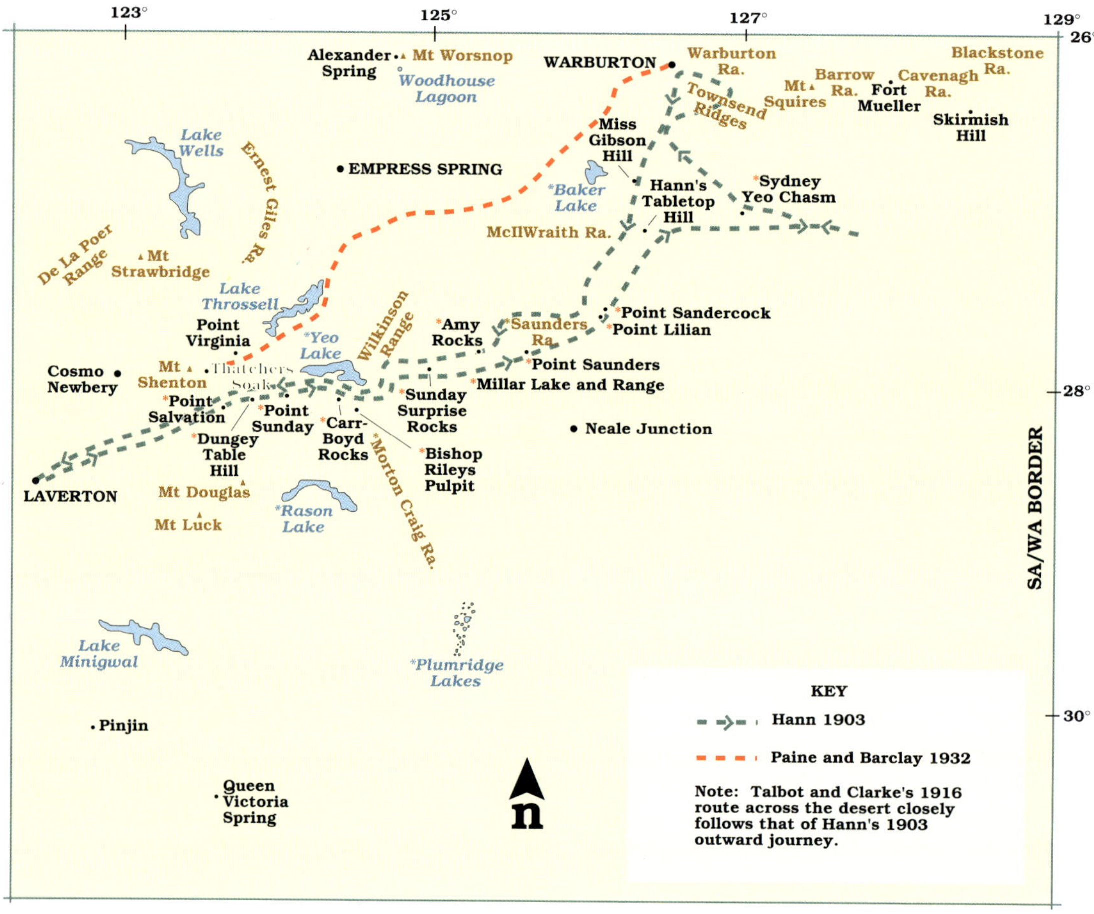

Exploration routes of Frank Hann 1903 and Paine and Barclay 1932. Selected geographical features named by Frank Hann are marked by an asterisk. This map also shows the location of some of the sites visited by other explorers of the north-western Great Victoria Desert and beyond.

Mike Donaldson

Parallel longitudinal dunefields are found near Point Saunders.

Mike Donaldson

LEFT: The rare claypans of Millar Lake, near the Millar Range, are clearly visible from 10,000 metres on the Perth-Alice Springs flight path.

Mike Donaldson

RIGHT: This dead mulga tree north of Lake Rason was blazed by Frank Hann on 7 September 1907. The blazed tree was relocated by geologist Mike Donaldson in 1991. The blaze shows the inscriptions 'FH' over 'T'. The letters 'HMc' were also cut below the T, but are now barely discernible. The initials are those of Frank Hann, his Aboriginal companion Talbot, and Henry McCarthy.

BELOW: A sea of mulga gives way to saltbush and grasses adjacent to the low breakaways of the Saunders Range.

Mike Donaldson

route took him via Point Sunday, the Yeo Hills, Lily Rock Hole, Millar Range, Point Sandercock and Sydney Yeo Chasm. He returned to Laverton and reported that a practicable stock route to the Warburton Range had been discovered.

In late 1903 to early 1904, Hann travelled from Laverton, via a similar path across the Great Victoria Desert and along the desert's fringing northern range country, to Oodnadatta and back again. He then ventured from Laverton across the north-western Great Victoria Desert to the Petermann Range in the Northern Territory between June and October 1904.

Hann made several trips from Laverton to the area around Cosmo Newbery and Point Salvation (near Yamarna) in 1905 and 1906. He took a paid hand, Syd Giles, with him during this period. He was accompanied by an Aboriginal boy named Talbot throughout all his trips across the Great Victoria Desert.

Hann later travelled from Laverton across the north-western desert to the South Australian border and back between February and July 1906. He again journeyed from Laverton to Oodnadatta via the desert from December 1906 to February 1907, and returned in June–September 1907. He then explored around the western and southern fringes of the Great Victoria Desert during late 1907 and 1908, visiting Pinjin, the Ernest Giles Range, and the Queen Victoria Spring area before making a final trip to the Warburton Range. He was accompanied by Talbot and a prospector, H. Plumridge, on his journey to the Queen Victoria Spring area in July–August 1908. Since Giles' initial discovery, Queen Victoria Spring had been visited regularly by prospectors including Fred Newman (1890), Thomas Tierney (1894) and J.E. Tregurtha (1899) as well as by the geologist Charles Gibson (1908).[28]

The original purpose of Frank Hann's exploration trips in and around the Great Victoria Desert was to search for new pastoral lands. However, he was also an accomplished prospector, and was responsible for finding the first copper deposits in the Warburton area. This discovery prompted the Western Australian Government to undertake a geological survey between Laverton and the South Australian border in 1916 (see page 132).

Hann prospected around Laverton for some years after 1908, but died a pauper in August 1921. He was well known and widely respected for his exploration feats, and his death was marked by articles in papers all over the country. He named more geographical features in Western Australia than did any other single person. These number more than 100 in the Great Victoria Desert area alone, and include Lake Rason, the Yeo Lake and Hills, Plumridge Lakes, Morton Craig Range, Millar Lake and Range, Point Lilian, Point Salvation, Point Sandercock, Point Sunday, Carr-Boyd Rocks, Amy Rocks, Point Saunders and the Saunders Range, to mention but a few. The Hann River and Mount Hann in the Kimberley, the Hann Tabletop along the Connie Sue Highway just outside the Great Victoria Desert, and the Frank Hann National Park south-west of Norseman are all named in his honour. There is a street in a new subdivision of Laverton named after him, and Hann's Camp about 10 kilometres east of Laverton is marked on many modern topographical maps.

Most of Frank Hann's diaries and manuscripts are now housed in the Battye Library in Perth.[44] A facsimile edition of his exploration diaries is currently being compiled and edited by Mike Donaldson and Ian Elliot.[45]

Frank George

A South Australian Government prospector, Frank Rees George, undertook a prospecting expedition north-west of Ooldea in 1904.[46] The expedition was initially suggested to the South Australian Government by Richard Maurice through the RGSA. Maurice was unable to join the expedition due to his impending return to England, but supplied 13 camels and much of the party's general equipment.[47]

The objective of the expedition was to prospect the country north of the Nullarbor Plain to around latitude 28°45', and west of longitude 132° to the State border. Particular attention was to be given to the area's saltlakes, as Maurice had previously reported finding a small quantity of gold on a saltpan some 40 kilometres south of Waldana. Expedition members included George, a fellow prospector A.E. Treloar, camel-driver Robert Scott and Aboriginal guide Mungena.

The party left Fowlers Bay on 6 July and headed north to Pidinga, Ooldea and Paling rockhole. After travelling around 90 kilometres west of Paling, a small native dam called Arndinga was found just east of a large saltlake. George described Arndinga as elliptical in shape, 7 metres long, 5.5 metres wide and 30–45 cm deep. There were some low cliffs on the boundary of the claypan in which the dam was situated. George named the large saltpan Lake Maurice, after the man 'who has done so much towards the exploration of this back country'.[46]

A westerly course was then taken direct to Waldana. Its waters were flowing at around 1100 litres per day during their visit. George left Scott and Mungena at Waldana and took Treloar on a flying visit west to Boundary Dam. He noted that Giles had fixed the dam's position approximately 20 kilometres too far west.[46]

The party returned to Waldana and then headed south to Churina and Bringyna native wells. Both wells were situated on small loam depressions surrounded by high sandhills, 'the water being caught and conserved in the sand by a bed of porcelainised sandstone'.[46] Churina had a plentiful supply of water at a depth of 1.2 metres and was flowing at a rate of 70 litres per hour.

George then examined the country around Wyola Lakes and proceeded south-west towards the northern edge of the Nullarbor Plain. Yet another dam, Midgeninga, was found on the western end of a small saltlake. Muckera and Choolkooning rockholes were visited, but both contained little or no water. A northerly course was then adopted. A dry Chilbinga native dam was located on a narrow strip of low-lying gypseous country. The camels were now extremely thirsty and the party proceeded back to Churina and Bringyna native wells. The flow of water at Churina had greatly diminished. The well had to be dug to a depth greater than 3 metres before water was met.

Two days were spent examining the lakes near where Maurice had reported finding gold. After an unsuccessful search, George set a course direct for Ooldea and then to Pidinga, which was reached on 8 October. He left Fowlers Bay by ship 20 days later and returned to Adelaide on 6 November. He described the mineral potential of the area as follows:

> *No outcrops of metalliferous rocks were seen, nor was there any country discovered that offered the slightest inducement to stop and prospect; and I am of the opinion that none exists in the region under notice.*[46]

David Pearson

Mike Donaldson

ABOVE LEFT: This rock carving in the Townsend Ridges, south of the Warburton Range, features the names of Neal and James East, as well as Frank Hann and his Aboriginal assistant, Talbot.

LEFT: Henry Paine and Hugh Barclay placed a series of numbered posts along their route between Laverton and Warburton in 1932. Peg PB55 (pictured) was positioned at Gnamal rockhole near Lake Throssell.

In 1919, geologist Henry Talbot described Streich Mound as the highest sandhill in Western Australia. This landmark was named after Victor Streich, the geologist with the 1891–2 Elder Expedition.

David Pearson

Frank George died two years later on an expedition to the Petermann and Treur Ranges in the Northern Territory.

East Brothers

Neal and James East undertook a prospecting trip from Laverton to the Warburton Range in 1905.[48,49] The Canadian-born brothers, together with an Afghan camel driver, Miski, and six camels, left Laverton on 26 June. They initially travelled south-east into the Great Victoria Desert via Mt Luck and Mt Douglas, in an attempt to locate possible auriferous country reported by Victor Streich after the abandonment of the Elder Expedition.[28] Following a thorough but unsuccessful search, the brothers headed north across the desert's spinifex and sandhill country and eventually intersected the former tracks of Frank Hann. They then took a north-easterly course to Lilian Creek (on the northern side of the Townsend Ridges) across dry spinifex country with occasional breakaways. They continued to prospect through the Barrow and Warburton Ranges and eventually returned via the Great Victoria Desert to Laverton on 27 September. A plaque commemorating the East Brothers expedition has been erected in Laverton by family members, with the help of the local council.[49]

Talbot and Clarke

Two Western Australian Government geologists, Henry Talbot and Edward Clarke, conducted a reconnaissance of the country between Laverton and the South Australian border in 1916.[50] Their primary aim was to investigate potential mineral-bearing country in the Warburton Range, first reported by Frank Hann in 1903.

The two men, accompanied by four camp assistants and 16 camels, left Laverton on 29 June and headed east-north-east across the Great Victoria Desert. Their route, which followed closely that of Hann's, took them via Point Salvation, Carr-Boyd Rocks (near Yeo Lake and the Morton Craig Range), Point Lilian (in the Sherriff Range) and the Hann Breakaways, to Lilian Creek. A depot was established at the creek, from which three flying trips were made in late August–September to examine the ranges further north and east.

The party was attacked by local Aboriginal people near Mt Gosse on 10 September. One of the men, J.W. Johnson, was severely wounded, necessitating the early return of the expedition across the Great Victoria Desert to Laverton and curtailing field work by a month.

The expedition's report contains detailed accounts of the physiography of the Great Victoria Desert's sandplain, sandridge and breakaway country. Meticulous descriptions of the location of water sources across the north-western desert (mainly rockholes in the breakaway country) are also given. The report pays tribute to the work of Hann by stating:

> *Hann has visited nearly every hill in this region, and we found very few waters that were missed by him... Hann has done exceedingly useful work in this district, and his mapping of the physical features compares favourably with that done in other parts of the State by more elaborately equipped explorers.*[50]

Talbot returned to a different part of the desert in 1919, this time conducting a geological survey of the country north of the Transcontinental Railway Line and east of Randall's siding.[51] He visited Queen Victoria Spring during this survey, and relocated Streich Mound, which he described as 'the highest sandhill that I have seen in the State'.[51]

A Flurry of Activity in the Desert

Many small expeditions were mounted in and around the north-western Great Victoria Desert from the late 1910s to early 1930s as interest in prospecting (in particular) continued to grow. Unfortunately detailed information on many of these ventures is not readily available. Nonetheless, names like Tregurtha, Slavin and Alexander, Byers, Cable, Hazlett (see next section) and Gall, to mention but a few, feature importantly in this period of the desert's exploration history.[52]

Paine and Barclay

Western Australian surveyors Harry Paine and Hugh Barclay conducted a detailed traverse of the country between Laverton and the Warburton Range in 1932.[53] They positioned a series of numbered posts (marked 'PB') along their route and erected cairns on all major physical features. They also obtained information of native water sources (including capacity, permanency and Aboriginal names), made a botanical collection, and assessed the pastoral potential of the country over which they travelled.

Paine and Barclay were accompanied by a Western Australian Government geologist, F.C. Forman, whose task was to report on the mineral possibilities of the area and six other assistants (including two mechanics). A local Aboriginal man named Jacob also followed the party and provided Paine with the native names of watering places. Two motor trucks (a Ford and Chevrolet) and seven camels were used to transport petrol and provisions.

The party assembled at Laverton on 30 May 1931. They travelled via Cosmo Newbery and reached the starting point for their survey, Thatchers Soak (near Mt Shenton), on 4 June. To avoid the difficult sandhill country encountered by Talbot and Clarke during their 1916 geological reconnaissance, Paine and Barclay selected a more northerly and direct route to the Warburton area. This route had been used formerly by a dogger, Sam Hazlett, who was hoping to take up pastoral country near Warburton. Hazlett knew this country well and is believed to have made as many as 17 trips across the desert between 1904 and 1941.[54]

Paine and Barclay's party reached Minnie Creek by 6 June. Here they marked a soak with a mile peg numbered 'PB16', erected a cairn (HP30) on a nearby hill, and established a latitude station.

The surveyors followed a set routine as they progressed north-east across the Great Victoria Desert. Barclay continued the traverse forward from rockhole to rockhole, while Paine attended to the party's transport and supplies. Most of the rockholes were small, but contained plentiful water due to the good rainfall experienced that year. Paine considered that Buldyu Soak (PB84) near Lake Throssell was the only water source of any permanency between Minnie Creek and Elder Creek (near the Warburton Range).

The party met well-known author and pioneer outback motorist, Michael Terry, during mid-morning on 11 July.[53] Terry, who had also travelled along Hazlett's route in his truck, was returning from the Warburton Range area to Laverton. He stayed at Paine's camp for the remainder of the day and enjoyed an evening meal of 'fresh roast turkey'.[52]

Paine's team encountered Sam Hazlett and one of his sons on 29 July. The Hazletts were returning to Laverton and were accompanied by the Rundle Brothers, who had travelled from Karoonie (on the Transcontinental Railway Line near Kalgoorlie) to the Warburton Range, where they had spent time prospecting.

Paine and Barclay continued north-east and reached the Warburton area on 14 August. A numbered post (PB256) was inserted at the junction of the Elder and Hughes Creeks, near where Hazlett had sunk a well. Survey work continued further east to Lightning Rock during September. The party returned across the Great Victoria Desert the following month and arrived at Laverton on 14 October. Three days later, Paine and Barclay were back at the Western Australian Department of Land and Survey's head office in Perth.

Paine reported that the land between Thatchers Soak and Buldyu Soak was fair pastoral country, but from Buldyu to the Elder Creek, spinifex predominated and water was limited to small rockholes in low breakaways. Similar to Hann and Talbot, he concluded pastoralism would be a risky enterprise due to the extreme irregularity of the desert's rainfall and the great distance from any railhead. With the coming of the motor vehicle, the well-marked route of Paine and Barclay became an important road link between Laverton and Warburton. Today, the main road between these two locations follows closely the original path of the two Western Australian surveyors.

Concluding Remarks

The Great Victoria Desert had been the traditional home of Aboriginal people for thousands of years prior to the 1850s. However, as has been described in this chapter, the desert was subjected to a remarkably intensive period of European visitation between the 1850s and 1930s — by explorers, scientific parties, surveyors, geologists, prospectors and others such as doggers and railway workers (see Chapter 8). Concerted attempts were made to map and document the desert's features, understand its natural and cultural heritage, find new stock routes to open up pastoralism, and ascertain the area's mineral wealth. The latter two endeavours proved to be in vain. The poor auriferous nature of the Great Victoria Desert and its inability to sustain pastoralism had been firmly established by the 1930s. This exploration phase had nonetheless changed the face of the Great Victoria Desert forever. The consequences of this contact period upon the desert's Aboriginal people had now begun to emerge.

Chapter 8

RAILWAYS AND MISSIONS

The Contact Years

The exploration phase, described in Chapter 7, sowed the seeds of initial contact between Aboriginal people of the Great Victoria Desert and Europeans. White settlements began to spring up around the fringes of the desert. Access to the southern part of the desert was further opened up by the establishment of a major rail link between South Australia and Western Australia.

Many groups of Aboriginal people began gradually to leave the desert, as they became attracted increasingly by the range of European foods and material goods available at

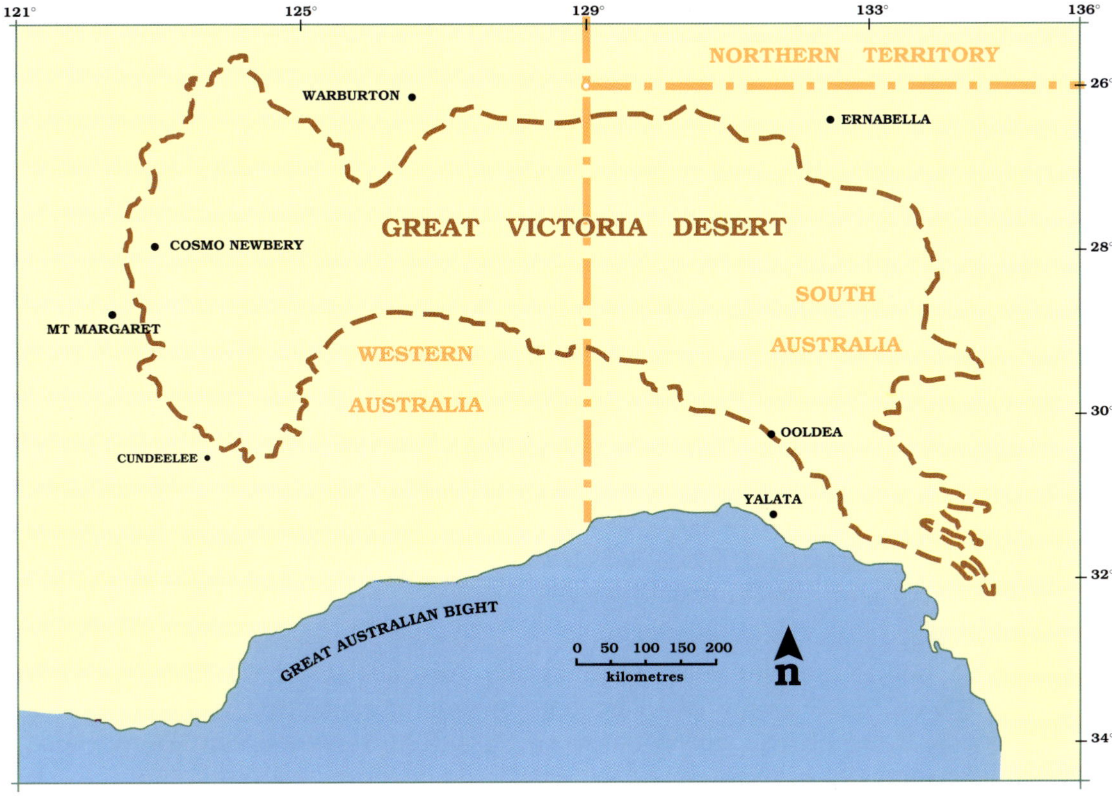

Location of Christian-based missions in and around the Great Victoria Desert.

settlements and by the opportunity to earn money in exchange for work. This contact phase created a new wave of problems for the desert's original inhabitants, including exploitation of labour, sexual abuse of Aboriginal women, exposure to diseases and alcohol, and dispossession from their traditional lands.

Government policies toward Aboriginal people underwent a series of changes during this period. In the late 1910s, policy focussed on 'protecting' Aboriginal communities living in the Central Australian region. The South Australian and Western Australian State Governments and the Federal Government were lobbied to create a 16.8-million hectare Central Reserve area 'for the use and benefit of Aboriginal people'.[1] Reserves were subsequently proclaimed in the Northern Territory (1920), Western Australia (1918) and South Australia (1921), the latter being known as the North West Aboriginal Reserve.[1] Additional areas of land were added to these original reserves over the ensuing years. Much of the northern Great Victoria Desert was encompassed within this initial reserve system.

Later government policies centred on providing welfare and attempting to assimilate Aboriginal people into the wider Australian community. In accordance with this latter philosophy, a number of Christian-based missions were founded around the edges of the Great Victoria Desert and beyond from the 1920s to the 1950s, at locations such as Ooldea, Cundeelee, Mount Margaret, the Warburton Range, Cosmo Newbery and Ernabella. These missions endeavoured to act as a buffer between Aboriginal people and the pitfalls of white contact by encouraging a gradual assimilation toward European lifestyles and by attempting to introduce the desert people to Christianity. Despite good intentions, the mission era became characterised by an increased dependence on European rations and, in some cases, Christian teachings were promulgated at the expense of traditional law and custom. Most missions became grossly overcrowded. Frustration, conflict and tension developed among Aboriginal residents as living conditions deteriorated, employment opportunities dwindled and pressure on basic resources increased.

The following description of the contact years at different locations around the desert, most notably at Ooldea, provides a powerful illustration of the consequences of this period of history up until the 1960s–1970s.

This description ends (for now) at the point where the focus of Federal Government Aboriginal policy in the 1970s changed from assimilation to self-determination and self-management. Under these new policies, Aboriginal people living at missions around the desert were encouraged to form incorporated community councils and manage their own affairs. The subsequent events that have reshaped and revitalised life for the Great Victoria Desert's Aboriginal people are discussed in Chapter 11.

Ooldea

From 'Desert People' to 'Railway Natives'

The South Australian Government surveyor J.S. Stewart visited Ooldea in 1901 while examining country from Tarcoola to the Western Australian border. He reported that Ooldea Soak had the potential to become an important source of water for the construction of an east-west railway line linking South Australia and Western Australia. A two-year survey of a proposed rail route between Port Augusta and Kalgoorlie was undertaken in 1907 using Commonwealth Government funds. Construction of the

Mortlock Library of South Australiana SSL: M: B45287/3

Mortlock Library of South Australiana SSL: M: B45287/3

ABOVE: Ooldea was a busy railway siding during its hey day in the late 1910s to early 1920s. None of the buildings in this picture remains standing today.

RIGHT: A closer view of the Ooldea Siding.

Transcontinental Railway Line began on 14 September 1912, with separate teams working west from Port Augusta and east from Kalgoorlie. They met on 17 October 1917 at a point west of Ooldea.[2] The most difficult stretch of line was encountered between Barton and Ooldea, where the Ooldea sandhills were met and more than one million cubic yards of sand were excavated in an 18-month period.[3] The first overland train left Port Augusta for Perth on 22 October 1917.

The tiny siding of Ooldea, some 8 kilometres south of the soak, was originally named Deakin 'to fit in with the proposed chronological naming of sidings after the Prime Ministers of Australia'.[2] However, the name Ooldea was subsequently retained after considerable local protest. Deakin was relocated further west, between Reid and Hughes.

Ooldea Soak was declared a water reserve by the South Australian Government and control over the area was handed to the Commonwealth Railways. A bore was sunk to supply water for operations at the siding. The prominent South Australian naturalist Captain Samuel White visited the soak on several occasions during 1917 and recorded that a condensing plant had been erected at Ooldea to 'freshen' the quality of the bore water.[4] White also noted that the felling of timber to fuel the condenser, together with the trampling of vegetation by camels carting water from Ooldea Soak to the siding, had impacted heavily on the local environment. He stated:

If [Maurice] could only see it now I am sure he would not know it with hundreds of reeking, sweating camels waiting for their loads of water for the railway siding of Ooldea. The country round has been trampled almost out of recognition.[4]

Anthony Bolam was posted to Ooldea as an operating porter in 1918. Bolam had joined the Victorian Railways in 1911 at the age of 18. He transferred to the Commonwealth Railways four years later and worked at various sidings along the east–west line before moving more permanently to Ooldea. During his seven-year stay at Ooldea, Bolam was promoted to station master and recorded observations on the Aboriginal and natural heritage of the region for his enchanting book *The Trans-Australian Wonderland*.[3]

Bolam stated that around 50 wells had been sunk at Ooldea Soak by the early 1920s. Pumping stations and a new pipeline supplied around 350,000 litres of water per week to the siding. Water was used to service the steam locomotives and to provide drinking water for the staff at Ooldea and other sidings down the line.[3]

According to Bolam:

The Soak is a wonderful asset to the Railways, because it is a permanent water supply in a country where a surface reservoir cannot be built owing to the sandy and porous nature of the soil; and there is no loss from evaporation, which in an open reservoir on this line is equal to 12 feet of water per annum.[3]

The construction of the Transcontinental Railway Line was nonetheless causing unforseen consequences in other areas. It began to signal the loss of autonomy of the desert's Aboriginal people. The native 'bush telegraph' told stories of the liberal availability of food, tobacco and alcohol at fettlers' camps and depots along the line. Aboriginal people who had never seen Europeans began to 'come out' of the Great Victoria Desert and congregate at Ooldea.[5-7] Many became exposed to introduced diseases against which they had no immunity. Sexual abuse of Aboriginal women by fettlers became rampant and many children of mixed descent were conceived. Begging was prevalent and people's self-esteem diminished as dependence on European material goods became entrenched.[6]

The Daisy Bates Era (1919–1934)

Daisy Bates arrived at Ooldea in September 1919. Her subsequent 16-year stay represented a remarkable, yet controversial era in the history of the region.

Daisy Bates (née Dwyer) was born in Ireland in October 1859.[8] She came to northern Queensland in 1884 and worked as a governess at station properties near Charters Towers and Nowra. She married twice within the space of 11 months.[9] Her first husband was Edwin Henry Murrant, the legendary 'Breaker Morant'. She then married John Bates and shortly after had a son. Recent research into this phase of Daisy Bates' life suggests that her first marriage was never annulled and that she kept this marriage a closely guarded secret.[8] She left her second husband and young son to return to England where she worked as a London journalist, but returned to Australia in 1899 to investigate for her newspaper allegations of cruelty to Aboriginal people.[7]

Bates' interest in Aboriginal people blossomed during subsequent work for the Western Australian Government. She prepared a detailed manuscript entitled *The native tribes of Western Australia*, but the State Government of the day refused to fund its publication. The manuscript lay dormant in the archives of the National Library in Canberra until its recent release in book form in 1985.[9] Bates was later employed as Protector of Aborigines at Eucla from 1912 to 1914. She then worked in the Fowlers Bay area before arriving at Ooldea.

Her initial accommodation was a tent pitched on the south side of the railway line.[10] At the time there were several hundred Aboriginal people living near the siding, most of whom were 'begging from the train at every stopping place...and already ravaged by disease'.[10] Her base camp was soon moved to a sandy gully around 2 kilometres north of the siding. It comprised a 2.4-metre by 3-metre tent, an upturned 2000-litre tank which functioned as a library, and a bough-shed storehouse which held supplies and numerous Aboriginal artefacts that were entrusted to her keeping. An observatory, described as a rickety bough shade with a crude bush ladder leading up to a leafy roof, was built on the crest of a nearby hill.[10] A windbreak of mulga branches surrounded her camp and an entrance 'gate' was made from clumps of acacia bushes. Aboriginal people were never invited into her domain during her 16-year stay. All exchanges of provisions took place just outside the entrance to the windbreak.[10] Her northern campsite has recently been the subject of a detailed anthropological study.[8]

Daisy Bates' self-appointed aims in coming to Ooldea were to provide Aboriginal people with food, clothing and simple medicines, to actively discourage contact between Aboriginal women and railway workers, and to 'explain the white man's ways and white man's laws'.[10] The South Australian Government refused to officially support her work, and subsequent letters written by Bates requesting assistance from the Prime

Daisy Bates, pictured outside her tent, was a self-appointed guardian of Ooldea's Aboriginal people from 1919 to 1934.

Anthony Bolam

Minister fell on deaf ears. Within a year, Bates was forced to sell her freehold residential estate overlooking the Swan River in Perth to finance her work. She also gained modest income over the ensuing years from 'the earnings of her pen', contributing numerous articles on her work to newspapers both in England and Australia.[11]

An epidemic of sandy blight (an eye irritation) broke out at Ooldea in November 1920. Bates became afflicted by the condition, which necessitated a visit to an eye specialist in Perth to prevent permanent blindness.[10] Dr Herbert Basedow visited Ooldea during the same year, as part of a South Australian Government-sponsored medical relief expedition.[8] He made a damning report on the health and general well-being of Aboriginal people living at Ooldea and was highly critical of Bates. He states:

> *At Ooldea, over a hundred people were detained for the simple reason, it appeared, of serving the ambitious aims of Mrs Daisy Bates.*[12]

Bates' presence was viewed as encouraging Aboriginal people to remain at Ooldea where they had become dependent on charity.[8] Attempts to discourage Aboriginal use of 'free' transport on open cattle trucks or on the 'tea and sugar' train, which provided regular provisions to railway employees, had also failed.[7]

Basedow recommended that Aboriginal people should not be allowed to camp within 16 kilometres of the railway line. Bates offered to relocate Ooldea's Aboriginal community at either Waldanna Well or Punthanna on several occasions, to ensure 'the natives...[could] lead their own natural lives without clothing and without cunning'.[10] Her offers were rebuked and the Chief Protector of Aborigines ordered her to leave Ooldea. Bates steadfastly refused and thereafter distanced herself from any further correspondence with State authorities.

In 1922, two bores were sunk at Ooldea Soak which yielded highly brackish water. Bates stated:

> *It was the beginning of the end for this magical Yul dilgabbi [Ooldea water] that had not failed in hundreds, perhaps thousands of generations.*[10]

She believed that the soak's clay base had been penetrated, which allowed the water reservoir to simply drain away. The official government viewpoint suggested that over-exploitation of Ooldea Soak, together with problems of increasing salinity, caused it to fail.[8] The soak was formally abandoned by the railways in October 1926. Water was subsequently railed from Kingoonya, 350 kilometres to the east.[8,10] The demise of Ooldea Soak and the deterioration of its water quality are seen by some as a poignant symbol of the degradation of Aboriginal culture which occurred at the time.[6]

To make matters more difficult, the destruction of Ooldea Soak also coincided with the beginning of an eight-year drought. Supplies of native foods dwindled and 'rain-songs were in vain'.[10] Raging sandstorms and oppressive summer temperatures made life at Ooldea close to unbearable. Bates' personal funds were low, but they received an unexpected boost when cheques for £7 10s and £60 were given by an American and Adelaide University, respectively.[10] Most of the money was immediately allocated to replenishing food supplies. The drought conditions worsened during 1928 and 1929 when Bates remarked:

> *fruits and berries had shrivelled... malleehens... had disappeared from the valleys, and the white man's rabbits were rarely to be seen on the sandhills they had infested in their millions.*[10]

A fire raged close to her camp on Christmas Eve 1929. Bates' precious manuscripts were placed in boxes and buried in a 2-metre deep pit, while a 20-metre fire-break was cleared around the camp. The fire burnt itself out within a short distance of her tent. The drought broke shortly after with a 'whirling mass of sand and wind and rain'. Bates recalled:

There were many gusty showers; and after the parched years, a beautiful vision. Green returned to the earth and the world was filled with the sweet fresh scent of herbage.[10]

Daisy Bates left Ooldea during 1934 and returned to Adelaide. During her 16-year stay, Bates was known to Aboriginal people as Kabbarli, or grandmother.[10,13] She became far more than just a provider of food and medicines. She claims that she was allowed to participate in selected ceremonies and became custodian of sacred religious objects. A wealth of information was collected on Aboriginal life, social structure, language and religious beliefs.

My voluminous notes have been scribbled anyhow and anywhere, on white paper and brown, diaries and notebooks and illegible and unintelligible to any same me.[10]

These notes later formed the basis for part of a book which Bates published, entitled *The passing of the Aborigines*.[10]

Daisy Bates has been viewed in sharply contrasting ways through the eyes of white Australians. Her uncompromising, indefatigable demands for support incurred the wrath of government authorities. Her persistent, ambitious struggle for official recognition as a pioneering anthropologist and Protector of Aborigines at Ooldea met with cold resistance. The unorthodox opinions she expressed (particularly on cannibalism among Aboriginal people) and her literary penchant for sensationalism caused her to be labelled by some as an irresponsible eccentric and a fraud.[11]

To others who read her articles or caught fleeting glimpses of her as they passed Ooldea by train, she was a dynamic and colourful personality with inexhaustible energy, an object of wonder and intrigue. Hill described her as 'the most remarkable woman in Australia's first two centuries'.[13] She was eventually honoured for her work by being awarded the Order of the British Empire.[10]

Daisy Bates was unquestionably a paradoxical character and recent research has thrown much of her early life into doubt. On the one hand, she portrayed the image of being very Victorian and upper class. She was always impeccably dressed and her long, flowing skirts, white gloves and wide-brimmed hats, all from an out-dated era, became one of her trademarks. Yet despite her apparent pedigree, she chose to endure a life of considerable hardship, ill-health and personal deprivation at an isolated desert outpost to champion the cause of black Australians.

Perhaps final judgement on Daisy Bates should be left to those Aboriginal people who are still living and remember her days at Ooldea. To them, she is generally thought of with fondness and affection.[11]

The United Aborigines Mission at Ooldea (1933–1951)

In the early 1930s, a Methodist Home missionary from Tarcoola visited the Council of the United Aborigines Mission (UAM) in Adelaide. He implored the Council to establish a mission at Ooldea Soak to act as a buffer between the Aboriginal people and the temptations that awaited them along the railway line. Fifty-six-year-old Annie Lock, who had spent 30 years working among Aboriginal people in four States, and a blind companion, Mary Battams, visited Ooldea Soak by train to survey the situation first hand. Lock's report recommended that a mission should be opened immediately.[14]

On 12 July 1933, Annie Lock and an associate, Miss Marshall, left Balaklava near Adelaide on an 800-kilometre overland drive by horse and buggy to Ooldea Soak. A small railway hut, a derelict railway cottage and some rusty sheets of galvanised iron were all that remained at the soak. The cottage was soon renovated and became home to Lock.

The mission received full support from the South Australian Government. Within two months, the Chief Protector of Aborigines was providing Lock with rations of tea, sugar and flour to distribute among Ooldea's Aboriginal people. Annie Lock's role in supplying rations was viewed by the South Australian Government as a last-ditch attempt to keep the people away from the railway line.[7] Not surprisingly, a major conflict of interest had developed between Daisy Bates and Annie Lock. Bates, who had fought an unsuccessful battle for recognition by State authorities and largely self-funded her work, reluctantly left Ooldea in 1934.[7,8]

Around 200 Aboriginal people were living at Ooldea by August 1934. Elsie Tyler joined the mission as schoolmistress. Initially, young Aboriginal boys and girls were assembled for school in a semicircle on the sand. Some months passed before a brush school, equipped with seats and desks, was built and children worked with individual blackboards and chalk. Reverend J.H. Sexton, from the Aborigines' Friends' Association (AFA), visited Ooldea during 1935 and was glowing in his praise for Lock's efforts. He stated:

> *With an intensely practical mind her program among the natives was well-arranged, and gave us great satisfaction... The situation for natives at Ooldea is ideal. They are unspoilt. They are too remote for the unprincipled white man to reach them to despoil their race.*[14]

Soon after Reverend Sexton's departure, a measles epidemic caused considerable suffering among the Aboriginal people camped at Ooldea Soak. The heavy burden of the epidemic took its toll on Annie Lock's health and she left Ooldea after three years' service.[14]

Harry Green, and his wife Marion, left Adelaide in April 1936 to replace Annie Lock at the mission. Green had conducted previous missionary work in Central and Western Australia. Once established at Ooldea, the couple set about increasing the availability of rations and providing better clothing for the native people.

Language difficulties were proving a barrier between the Greens and the Aboriginal people living at Ooldea. With the help of several younger Aboriginal men who knew some English, more than 20 Bible stories were translated into native dialect.

In May 1938, control of Ooldea Soak was transferred from the Commonwealth Railways to the South Australian Government, the former body no longer having any use for the soak. The State Government immediately set aside an area of approximately half a million hectares as the Ooldea Aboriginal Reserve.[5,8]

An ambitious building program began, using old railway sleepers which had been stacked and left in a large pile some 25 kilometres from Ooldea. An old buckboard was used to laboriously bring four sleepers at a time to the mission. The later purchase of an old 2-ton chevrolet increased cartage to 25. The sleepers were used to build a guest room, motor shed, shelter shed, bathroom, school and church, mission house, and goat's pen.

Several generous donations enabled a dormitory for 60 children to be constructed from entirely new building materials sent from Adelaide.

Life at the mission was busy. Women and young girls were taught basic domestic duties such as sewing, knitting, washing, ironing and cleaning.[6] Younger Aboriginal men were employed in a variety of tasks including tending the vegetable garden and fruit trees, milking and shepherding the mission's 300 goats, operating a sand-scoop to clear drift sand from the buildings, baking bread, carting water, cutting firewood, sawing sleepers and collecting mail from the siding.[14] Some men hunted animals for bounty

Mark Shephard

LEFT: The sandhills surrounding Ooldea Soak contain a high density of artefact scatter and indicate the traditional importance of the soak to Aboriginal people.

BELOW: Up to 500 Aboriginal people gathered in the sandhills at Ooldea Soak for traditional ceremonies during the 1940s.

Mark Shephard

while others carved implements that could be bartered or sold to train passengers. Rations of flour, sugar, tea, potatoes, onions and dripping were distributed three times a week. As Brady explains, Aboriginal people living at Ooldea Mission were 'psychologically addicted to European food, caught up in a particular rhythm of life, and involved in a new social relationship'.[7]

While population movements in and out of the desert continued through the 1940s, more and more people remained at Ooldea on a semipermanent basis. A constant battle was fought to keep the people occupied. A number of married couples were sent to a newly opened mission at Gerard on the River Murray in South Australia, while a contingent of young men were given the opportunity to work on farms and station properties well away from Ooldea.

Christian teaching was nonetheless the mission's foremost imperative. Aboriginal parents were encouraged to place their children in the mission's dormitories, where religious instruction and basic elements of education were taught. Attempts were also made to Christianise the people who maintained a more-or-less traditional lifestyle at camps in the sandhills around Ooldea Soak. By 1944, there were 44 children in the dormitories, while 24 remained in camps. Boys living at the dormitory were given a blue tie and girls wore a pink ribbon. Sixteen young people were baptised during 1944.[6]

Anthropologists Ronald and Catherine Berndt spent six months at Ooldea Soak in 1941. They were critical of the mission. The Berndts believed Ooldea's Aboriginal people were 'subjected to more or less constant pressure to renounce their primary Aboriginal association and become Christians'.[15] Nonetheless, traditional ceremonies continued to be practised through this period, and between 400 and 500 people gathered occasionally at these ceremonies.[15] The Berndts' six-month stay at Ooldea resulted in the publication of a series of detailed anthropological papers and books relating to the culture of the southern Great Victoria Desert's Aboriginal people.[15–19] They provide a valuable resource on this period of history.

The Closure of Ooldea Mission

Living conditions at Ooldea deteriorated significantly by the end of the 1940s. The local ecology was collapsing under the pressure of permanent habitation.[6] The sandhills surrounding Ooldea Soak had become almost completely denuded as a result of the impact of rabbits, goats and the cutting of timber for firewood. Resultant uncontrollable sand drifts threatened to envelop the mission's buildings, many of which were riddled with white ants. Water tanks were now badly corroded. There were still more than 300 Aboriginal people (including 50 dormitory children) living at Ooldea — their future appeared bleak.[6]

The Aboriginal Protection Board's 1949 Annual Report cast serious doubt on the future viability of Ooldea Mission.[7] In March 1951, the South Australian Government purchased Yalata Station, 200 kilometres west of Ceduna, with the intention of resettling Ooldea's Aboriginal population. There was immediate opposition to the purchase from the Lutheran Church, who claimed UAM's relocation would have adverse effects on their own mission at Koonibba.[20] The South Australian Government was petitioned to grant control of the new mission to the Lutherans. Neither the UAM nor the Lutheran Church would agree to joint management. The conflict was resolved following an internal dispute within the UAM which ultimately precipitated Ooldea's closure. The South Australian Branch of the UAM sought autonomy from the Federal Council and formed a breakaway group.[7,21,22] The federal body decided to withdraw its workers from Ooldea immediately. The entire missionary staff, led by Harry Green, left Ooldea on 24 June 1952.[6] The South Australian branch had no staff available to take over duties at Ooldea. The Protector of Aborigines therefore asked the Koonibba Mission to intercede at short notice and undertake the resettlement of Ooldea's Aboriginal people.[20,23] Hans Gaden, the station manager at Koonibba, received an urgent government telegram to 'take over Ooldea and Maralinga the following Monday am'.[6]

Aboriginal people at Ooldea were told they had to move, but were given no indication where they were being taken or for how long. As Brady states:

> *There can be no doubt that they had no intimation that a later return would not be possible, and certainly it was inconceivable that they could 'lose' their country.*[7]

Considerable trauma, confusion and heartache were suffered by Aboriginal people as a result of the abruptness of Ooldea's closure and their impending removal from the area.

Hans Gaden arrived at Ooldea with instructions to take the 50 mission children to Koonibba as an interim measure and to move the remaining 300 Aboriginal people to a temporary settlement at Yalata Station.[2] Only around 65 people (all Kokatha) were prepared to go with him.[6] They were taken to Ooldea siding and left overnight. Ian McLeod (a policeman from Cook) and Walter MacDougall (a native patrol officer—see Chapter 9), arrived the following morning to assist Gaden with the relocation.[6] The Aboriginal people were taken by truck to Ooldea Tank, 27 kilometres north-east of Yalata. Only six children were transferred to the Koonibba Mission.[20]

Some of the people remaining at Ooldea wanted to go west to Cundeelee. The majority, who were Pitjantjatjara, desired to return to Ernabella in the Musgrave Ranges. They caught the train to Tarcoola with the intention of heading north, but were intercepted by MacDougall and forced to return to Ooldea. Around 400 Pitjantjatjara people were relocated to Yalata over the following months.[6]

The Lutheran Church dismantled some of the buildings at Ooldea during 1954 and transported them by truck to Yalata.[5] Ooldea railway siding closed and its staff were transferred to Watson. Patrol Officer MacDougall removed all remaining Aboriginal items from the area. Ooldea's status as an Aboriginal Reserve was formally revoked on 9 December 1954 after a request from Prime Minister Menzies.[5] A year later, most of the Great Victoria Desert north of Ooldea was declared a Prohibited Area in preparation for impending British nuclear tests (see Chapter 9).[5] Although Ooldea was officially 2 kilometres outside the southern boundary of the Prohibited Area, any further visits to the region by Aboriginal people were actively discouraged.

The Ooldea era was over. The area nonetheless retains significant current interest for both European and Aboriginal people. Daisy Bates' campsite contains numerous artefacts of European origin and a management plan for its preservation has been recommended.[2] Although now completely engulfed by sand, there is also future potential for the excavation of some of Ooldea Mission's foundations.

Aboriginal people view Ooldea with a mixture of strong emotions, ranging from joy and pride to melancholy and regret.[2] Spiritual links with the area remain firm. Ooldea was the first point of contact with Europeans for many people. Those who were born or educated there retain a sentimental bond. Others remember Ooldea as their last camping ground on 'clean, red sand' before transferring to Yalata.[5]

Life at Yalata

The Lutheran Church took over official responsibility for Yalata on 1 July 1954.[20–24] The present-day site of Yalata at Tallowan Tank was selected a month later and work began on building a permanent settlement. Attempts were made to re-establish sheep on the station. The first baptism service was held in 1956.[20]

History was repeating itself within a decade, however. Problems with unemployment and the settlement's water supply were experienced. Some people had once again become dependent on the mission. Others attempted to sell artefacts to travellers on the nearby Eyre Highway but, in doing so, became exposed to the perils of alcohol.

In an attempt to rectify these problems, a system of 'big' and 'little' camps was established (and is still practised today).[22,24] Most of Yalata's 350 Aboriginal residents lived in traditional accommodation at 'big' camps, which were moved several times a year and serviced by mobile water tanks. Big camps were often as far as 20 kilometres from the main settlement. Aged and infirmed Aboriginal people resided in 'little camps' close to the settlement's central core of administrative buildings, church, school, store and hospital clinic

Many Aboriginal people became discontent with the 'dusty, grey earth' of Yalata and yearned to return to the 'clean, red sands' of the desert.[22] To quote Palmer:

> *The dissatisfaction people feel with Yalata is a consequence, so they say, of the earth and soil there which is grey and which in turn makes their hair grey (ie, makes them old) and causes them great worry. When they make this assertion, the Aborigines are expressing their resentment at the grey earth and are stating that their alienation is manifest in real physical terms... The attitudes of the Yalata people concerning their present place of residence is compounded into a single metaphorical statement.*[22]

Yalata was simply not part of the people's traditional lands and they were concerned and unsettled about the fate of 'their country'. As discussed in Chapter 6, the congregation of people at Yalata resulted in the emergence of a new cultural uniformity among its residents.[7,24]

The agreement between the South Australian Government and the Lutheran Church over control of Yalata expired on 30 June 1975.[20] Innovative Federal Government policy in the early 1970s sought to provide Aboriginal people with a degree of independence, and a philosophy of self-determination and self-management now replaced the Government's earlier policy of assimilation (see Chapter 11).[23] Both the mission staff of Yalata and the Board of Aboriginal Missions of the Lutheran Church of Australia encouraged the settlement's Aboriginal people to form their own body, Yalata Community Incorporated;[20] this new body took over self-administration of Yalata Reserve on 1 July 1975.[24] The title to the Reserve's 456,150 hectares of land was vested in the South Australian Lands Trust and leased to the incorporated Aboriginal body at a nominal rent.[22]

Cundeelee Mission

The problems caused by the construction of the Transcontinental Railway Line were not confined to the Ooldea region of South Australia. Aboriginal people from the southern

Mark Shephard

The Ooldea area contains many reminders of former European occupation.

Western Australian side of the desert also continued to drift towards the railway line and were exposed to problems similar to their South Australian counterparts. Cundeelee, a former government ration depot, was selected as the site for the Australian Aboriginal Evangelical Mission in 1950.[25,26] Cundeelee is located approximately 40 kilometres south-west of Queen Victoria Spring and north of the railway siding of Zanthus. Mr Telfer was appointed as the mission's first superintendent.

The mission initially consisted of a few tents and offered basic supplies of rations and medicines. A rapid influx of desert people occurred during the first two years of its operation, with 70 people arriving from South Australia in 1951 and another 60 from the Western Australian side of the desert in 1952.[26] More than 50 people walked from Yalata to Cundeelee for ceremonial purposes during the following year, but they were returned to South Australia by train.[26] The mission fought desperately for survival during these early years against a decided lack of government support, a shortage of water and the remoteness of its location.

In 1954, Albert Anderson led a 'contact' mission north-east from Cundeelee into the desert to seek out the relatives of Aboriginal people now living at the mission.[25] This was the first of around a dozen subsequent contact missions which took place over the next seven years in the vicinity of the Great Victoria Desert Nature Reserve. The trips were undertaken specifically at the behest of the Aboriginal people of Cundeelee, who wanted to be reunited with their relatives.[25] In 1958, the then superintendent Bob Stewart was accompanied on one of these contact expeditions by a party of well-known naturalists including Vincent and Dominic Serventy and a young assistant, Harry Butler.[27,28] Government patrols conducted during the same period also made contact with the desert people. The patrols were part of the British nuclear weapons testing program and were carried out by Patrol Officer Robert MacCauley (see Chapter 9).[26]

Cundeelee was now viewed by the Australian Federal Government as an important contact point for the desert's Aboriginal people.[26] The mission was experiencing severe problems, however. The number of Aboriginal residents had grown to about 200. There were limited employment opportunities apart from small-scale artefact production and sandalwood gathering. The general health of the community was poor, and a measles epidemic in 1965 did little to improve their situation. A serious shortage of water remained a constant concern despite the sinking of numerous bores.

Cundeelee was proclaimed a C Class Aboriginal Reserve in June 1973. The Aboriginal people of Cundeelee formed their own corporation, Upurl Upurlila Ngurratja, in 1977 and the mission relinquished control of the Reserve.[25] As described in Chapter 11, there would be further movement for the people of Cundeelee before a return to their traditional lands would become possible.

Mount Margaret Mission

In the early 1920s a mission was established at Mount Margaret, just outside the western boundary of the Great Victoria Desert. Many of the mission's Aboriginal residents originated initially from the North-Eastern Goldfields region. However, during the 1930s and 1940s in particular, a number of people came out of the Great Victoria Desert to live at the mission. The Goldfields people were different linguistically from those inhabiting the desert country, but were still part of the Western Desert Culture Bloc.

On 22 April 1921, Rodolphe (Rod) Samuel Schenk travelled by train from Melbourne to Laverton.[29] Rod was a member of the Australian Aborigines Mission, which following an internal split in 1929 became known as the United Aborigines Mission (UAM). His initial intention was to conduct an itinerant mission among Aboriginal people in the North-Eastern Goldfields region of Western Australia.[30] After spending a short period in Laverton, he decided a permanent settlement or mission should be established where Aboriginal people from the region and the neighbouring Great Victoria Desert could become more productive.

Schenk obtained a lease for 80 hectares of land within Mount Margaret Common in January 1922. Around 50 Aboriginal people sat down at the mission by September. Schenk married Mysie Johnston in Melbourne during October and the couple returned to Mount Margaret to resume mission work.[29]

A number of industries and crafts were established at the mission over the next 30 years, including sandalwood gathering, gold mining, raffia spinning and weaving.

The mission's first school was started on 6 September 1926 and the Graham Homes dormitory opened two years later. Mrs Bennett took over as school mistress in the early 1930s and the standard of education at the mission improved greatly. The school's later achievements in numeracy and literacy were widely lauded and set Mount Margaret apart from other missions of the period.[30]

However, life at the mission was not without hardship or conflict. Schenk waged a long-running feud with Laverton's Pastoral Association over wages for Aboriginal people. From 1929 to 1934, the Association tried valiantly to have the mission closed and relocated well to the east of Laverton in 'useless spinifex country' of the Great Victoria Desert.[29] The commercialisation of industry at the mission also drew opposition from some church groups.[30]

Initial relations between Schenk and the Western Australian Chief Protector of Aborigines A.O. Neville were good. Neville visited the mission during the late 1920s and prepared a positive report on its activities and operation. Conflict developed between the pair during the 1930s and 1940s due to Neville's insistence on opening a government ration depot at Cosmo Newbery, his lack of support for a new mission in the Warburton Range area (see later), and his acceptance of the State Government's radical *Administration Act 1936*.[29]

Well-known anthropologist A.P. Elkin visited Mount Margaret for two weeks in late 1930. He described the mission's Christian teachings as too fundamentalist and was critical of other mission policies.[30] In contrast, two other noted anthropologists, Professor J.B. Birdsell and Dr Norman Tindale, journeyed to Mount Margaret in 1939 and were greatly impressed by the mission's 'well thought-out goals' and 'good administration'.[29]

Mount Margaret endured a long period of drought during the 1930s. The community was also ravaged by a series of disease epidemics: influenza in 1924, typhoid in 1935 and measles in 1939.[29] The mission's Aboriginal population continued to grow from 50 in 1922 to around 300 in 1930.

Schenk and his wife Mysie retired from Mount Margaret in 1954.[29] The mission had developed impressively under Schenk's superintendence, largely as a result of accomplishments in education and industry and the continuity of service among its staff.

The mission's new leaders encouraged the development of an Aboriginal Church and its own ministers. A Western Desert Bible School and Translation Centre was established at the mission by Wilf Douglas in 1957.[31] Mount Margaret became an important evangelical centre by the late 1960s/early 1970s. An incorporated Aboriginal body, known as the Aboriginal Movement for Outback Survival (AMOS), was formed and eventually assumed administrative control of Mount Margaret mission around 1976.[30]

Warburton Ranges Mission

The Schenk era was also responsible for the establishment of the Warburton Ranges Mission. The Warburton Range lies outside the boundaries of the Great Victoria Desert proper, but the mission's founding involved numerous expeditions back and forth across the desert. Once established, Warburton Ranges Mission also provided a haven to many Aboriginal people who formerly inhabited the northern part of the Great Victoria Desert.

Schenk received several reports from prospecting parties in the early 1930s that many Aboriginal people were living in the Great Victoria Desert between Mount Margaret and the Warburton Range area. He became determined to establish mission outposts around the desert fringe. Reg Williams, who had been a missionary at Mount Margaret in 1926, offered to select a site for a new mission but was later forced to withdraw his services.[29] Williams later became well known as the founder of RM Williams Clothing Company. Will Wade, a friend of Williams, was an eager replacement.[30]

Wade and Fred Jackson left Mount Margaret on 7 June 1933 and headed north-east through the Great Victoria Desert on an initial reconnaissance to the Central Range country.[29]

In July 1934, Wade and Jackson, accompanied by Harry Lupton, Lindsay Lovick and a new team of camels, retraced their steps from Mount Margaret across the desert to the range country. They reached the junction of the Elder and Hughes Creeks near the Warburton Range after five weeks' travel.[32,33] Lupton and Jackson remained there, while Wade and Lovick were led by a young Aboriginal boy north towards Sladen Waters in the Rawlinson Ranges. They camped with a party of 200 Aboriginal people around 160 kilometres north-east of the creek junction. The Aborigines took the men's bag of flour during the night in a deliberate attempt to force their return to the Warburton Range area and thereby avoid hostility which awaited them further north.[32,33] Lupton and Jackson had sunk a well during their absence and reached water at 5 metres. The site for a new mission had been found. The entire party then made a 16-day, 640-kilometre return journey to Mount Margaret across the Great Victoria Desert.

Wade, his wife Iris and their two children, Lupton and his wife Ethel, and Harry Green undertook a seven-week trek via the desert to the Warburton Range in October 1934.

The Wades stayed at the site over Christmas while the Luptons and Green returned to Mount Margaret. In May 1935 Schenk received tentative permission to establish a mission in the Warburton Range area from A.O. Neville, almost three years since the concept was first proposed.[29]

It was soon realised that the original mission site was subject to flash flooding.[32] The present-day location of Warburton, some 5 kilometres away, was selected as the mission's new base. The Wades left the Warburton Ranges Mission in 1944 after a decade of service, which saw the mission firmly established and a school started.[33] However, the school and dormitory system was not entirely successful because Aboriginal parents were determined to exercise traditional authority over their children. A new school opened in the early 1960s, the dormitory was closed, and children were left in the care of their families. Qualified teachers from the Western Australian Education Department also began teaching at the school around this time.[33]

The area around the Warburton Range was opened to mining in 1966. Mission staff were now outnumbered by government and mining people, and the Aboriginal population of the Warburton Ranges Mission had grown to nearly 400.[34] The mission attracted Aboriginal people from a 700-kilometre radius, including the northern Great Victoria Desert. They came to the mission for a variety of reasons. Some sought reliable food and water resources after several periods of drought in 1953–1956 and 1960–1966. Others came out of curiosity, in response to the urging of relatives revisiting their homelands. A number were relocated by native patrols (see Chapter 9).[34] The congregation of people at the mission led to a socially tense and volatile environment. A new settlement was established at Docker River in 1968 to relieve some of the stress of mission life. In 1973 the UAM relinquished control of the Warburton Ranges Mission and an incorporated Aboriginal community with an elected local council was formed to take over its management.[34]

Cosmo Newbery

After Warburton's establishment, Schenk had visions of starting another mission in the Great Victoria Desert between Mount Margaret and the range country.

Cosmo Newbery, a pastoral property around 110 kilometres north-east of Laverton, became available for sale in 1939. Schenk was keen to purchase the property, but the Chief Protector of Aborigines A.O. Neville pre-empted Schenk's plans and opened a government ration depot at Cosmo Newbery in January 1940.[29] Neville retired from the public service two months' later. Cosmo Newbery became a penal settlement for Aboriginal offenders in subsequent years.[29] In a twist of fate, Cosmo Newbery was handed over to the United Aborigines Mission in 1953 under the initial superintendence of Harry and Ethel Lupton.[29] The mission ran more than 1000 cattle and 2000 sheep on its property during the 1960s.[28] Control of the mission passed to the newly formed Cosmo Newbery Aboriginal Corporation in the late 1970s.

Ernabella Mission

Ernabella, like Warburton, is nestled in the range country to the north of the Great Victoria Desert. The history of the mission is given brief coverage because it provided a base for many of the former traditional inhabitants of the north-eastern Great Victoria Desert.

Len Beadell

A Presbyterian Mission was established at Ernabella in the Musgrave Ranges in 1937. Many Aboriginal people from the north-eastern Great Victoria Desert came to live and work at the Mission. The Mission's Church is pictured around 1952.

Adelaide surgeon Dr Charles Duguid visited the Musgrave Ranges in the north-west of South Australia in 1935. He was concerned by the exploitation of Aboriginal people and their drift towards white settlements further east.[35] On his return to Adelaide, Dr Duguid campaigned to the South Australian Government to establish a buffer settlement in the region. Ernabella Mission was subsequently opened by the Presbyterian Church in 1937.[35,36] Aboriginal people at Ernabella were encouraged to continue their traditional lifestyle with no deliberate interference to their custom or law.[35,36] A mission school opened in 1940, but children remained with their families and were not placed in dormitories.[35] Mission workers attempted to learn traditional language.

A handicraft industry was commenced at the mission in 1948 and provided employment for Aboriginal women — this is still a successful industry today. Many of the men worked as stockmen on adjacent pastoral properties.[35]

Ernabella's Aboriginal population had risen to around 400 by the 1950s. The congregation of people at a centralised base again began to create problems. There was insufficient employment and pressure on resources such as water and firewood. These contributed to feelings of frustration among its residents.

A new South Australian Government settlement was opened at Musgrave Park (now Amata) in April 1961 to ease the problems being experienced at Ernabella. Fregon (Kaltjiti) was established 60 kilometres south-west of Ernabella during the same year. This latter initiative aimed to provide training and employment in cattle work, and to serve as a place of residence for people associated with the sandhill country of the Great Victoria Desert.[35,36] In the early to mid-1970s incorporated Aboriginal Community Councils were formed at Ernabella, Fregon and Amata, as well as at other settlements in areas such as Indulkana, Mimili (formerly Everard Park cattle station), Kenmore Park and Pipalyatjara.[35,36] The Pitjantjatjara Council was formed in 1976 to represent all communities throughout the region.[35]

The wheel was now beginning to turn full circle for the desert's Aboriginal people who lived at each of the missions described. Both Federal and State Governments and mission staff had sensed a growing frustration among the people and the need for change.[35] New federal policies of self-determination and self-management now enabled Aboriginal people, through their incorporated community council structures, to pursue land rights claims, return to their traditional lands, develop homelands or outstations, and revive aspects of cultural practices.[23] These activities will be described more fully in Chapter 11.

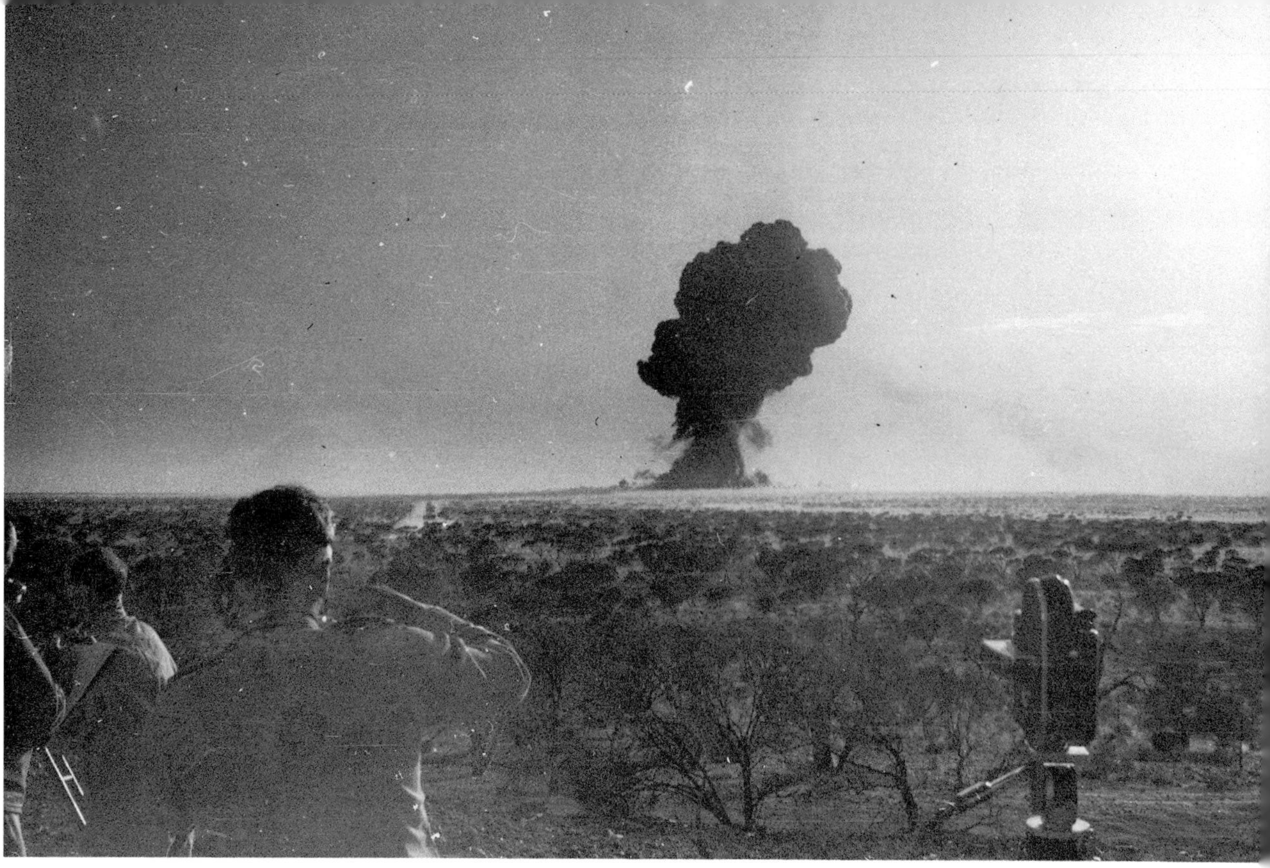

Len Beadell

ABOVE: Len Beadell took this photograph of the radioactive cloud of Totem 1, five seconds after its firing.

Len Beadell

ABOVE: The task of establishing a township at Emu was given to Len Beadell.

Chris Wenham

RIGHT: Totem 1 was detonated at Emu on 15 October 1953. A concrete plinth now marks the ground-zero site.

Chapter 9

ROCKETS AND ATOMIC TESTS

As well as railways and missions, there were other major developments during the 1940s to 1960s that changed the face of the Great Victoria Desert forever. These events included the building of a rocket range at Woomera and the commencement of British nuclear weapons testing on Australian soil.

Rockets Across the Desert

In 1947, the Australian and British Governments jointly established a rocket range at Woomera, 500 kilometres north-west of Adelaide and adjacent to the south-eastern Great Victoria Desert.[1] The British were keen to develop their own rockets to match or better the capabilities of the German V2 rocket that had been used to bomb London during the Second World War.[2] A large area of land, with an extension over the sea, was needed for a testing range. The close proximity of the Great Victoria Desert rendered it an obvious choice for rocket testing. The Woomera Prohibited Area was subsequently proclaimed. It included the area north and west of Woomera to the South Australian–Western Australian border, and north of the Transcontinental Railway Line to the then North West Aboriginal Reserve (now the Anangu Pitjantjatjara Lands). A second prohibited zone was declared in the north-west of Western Australia, with a 2400-kilometre extension into the Indian Ocean.[2]

Concern was expressed by sections of the general public over the joint venture about its potential to disrupt the lifestyle and endanger the lives of Aboriginal people.[3] As a result, a meeting of the Australian Committee on Guided Projectiles was held, and Dr Charles Duguid and Dr Donald Thomson were invited to the meeting to represent the Aboriginal cause.[3]

Walter MacDougall was subsequently appointed as Native Patrol Officer. His charter was to monitor the movements of Aboriginal people in the Great Victoria Desert and to keep them from entering the Prohibited Area. MacDougall, who was born in Melbourne in 1907, had spent a great deal of his life among Aboriginal people of Central Australia and was based at Ernabella for a number of years.[1] While well credentialled for the job, he was given a daunting task in carrying out ground surveillance duties which encompassed an area five times that of the United Kingdom. MacDougall's role as Native Patrol Officer continued well into the later British nuclear testing program.

Len Beadell, who had selected the site of Woomera in 1947, was given the task of constructing a network of roads across the Great Victoria Desert (as well as the Gibson and Great Sandy Deserts). The roads were to be used to provide access along the flight paths of the Woomera rockets. Beadell's story will be outlined more fully in Chapter 10.

RIGHT: Sunset over myall Acacia papyrocarpa *woodland, 1 kilometre south of the Totem 1 and 2 bombsites at Emu.*

BELOW RIGHT: Large aircraft were able to land on Emu claypan (pictured), but the high cost of airfreighting materials to the remote site was a major factor in relocating the British atomic testing program to Maralinga.

Radioactive Clouds Across the Desert

As well as guided-missile testing, the British Government was also keen to embark on a program of nuclear (atomic) weapons testing. On 19 February 1952, a joint Prime Ministerial statement announced that the United Kingdom would explode a nuclear device on Australian soil.[4] The first British atomic test in Australia was carried out in the Monte Bello Islands, 120 kilometres north-north-west of Dampier in Western Australia, on 3 October 1952.[4,5] The test was conducted on the Royal Navy ship HMS *Plym*, just off Trimouille Island, one of the Monte Bello group. Two further tests were undertaken on the islands during 1956.[4,5]

However, the British Government wanted to establish a more permanent ground-testing and weapons research centre on mainland Australia.[4,6] The vast open space of the Great Victoria Desert was again under consideration. An agreement in principle for mainland testing of atomic weapons at Emu, within the Woomera Prohibited Area, was sealed in December 1952.[4] This agreement was not publicly announced until 26 June 1953.[4] Although the British nuclear tests would have undoubtedly sealed the fate of Ooldea, the South Australian Government's decision to close Ooldea Mission one year earlier was not directly related to the impending test program.[7–9]

Atomic Tests at Emu

Emu is located in the heart of the eastern Great Victoria Desert, around 200 kilometres north of Ooldea and 280 kilometres west of Coober Pedy. The area surrounding Emu comprises a large claypan and gently undulating plains vegetated with mulga, myall and bluebush.

Two atomic explosions were conducted at Emu as part of Operation Totem. Totem 1 was detonated from a 30-metre tower on the morning of 15 October 1953, following a week of delays caused by bad weather and rain. Its explosive yield was 10 kilotons.[5,7] The radioactive cloud from Totem 1 rose 4500 metres after three minutes, remained intact, and drifted north-east.[7]

Several sets of aircraft monitored the cloud. A Canberra bomber flew through the cloud six minutes after the explosion to measure radiation levels.[4]. The cloud's path was tracked and air samples were taken by RAAF Lincoln and American B29 aircraft from Woomera and Richmond. The cloud crossed the Australian coastline near Townsville 50

Peter Canty

Mark Shephard

hours after the explosion.[7] Two Dakota aircraft from Woomera were also used for ground contamination surveys near Emu.[7]

Totem 2 was exploded from another 30-metre tower on the morning of 27 October. Its yield was 7–8 kilotons.[4,5] As a result of more turbulent wind conditions, the radioactive cloud of Totem 2 rose to around 6000 metres and was widely dispersed. Its fallout was therefore less concentrated. Air sampling and fallout surveys were again carried out by Australian aircraft but no planes flew through the cloud immediately after the explosion.[4]

The remoteness of Emu's location created major logistic problems, as a large amount of equipment and supplies had to be air-freighted to the area. A new location, which was both less geographically isolated and had better access, was sought.

Maralinga — Field of Thunder

During the Totem testing, Len Beadell had surveyed the area south and west of Emu and selected a potential new test site on Tietkins Plain. It was only 35 kilometres north-north-east of Watson on the Transcontinental Railway Line and could be readily serviced by train.

Like Emu, the site's terrain was generally flat and featureless, ideal for the establishment of a village and the detonation of bombs. Sir William Penney, Britain's chief nuclear scientist, inspected the site by air on 17 October 1953 and was impressed. The area was later given the name Maralinga, an Aboriginal word meaning 'thunder'.[4]

On 2 August 1954, Britain approached the Australian Government with a formal request to establish an atomic testing range at Maralinga.[4] The Australian Government agreed in principle to this request on 26 August 1954, but a Memorandum of Agreements between the two governments was not completed until 7 March 1956.[4,7] The agreement had been held up mainly by a compensation clause relating to who should bear the costs of any later claims for death or injury as a result of the tests. This clause would become 'the flashpoint of attention' some 30 years later.[1]

Maralinga township sprang up during late 1954 to early 1955 at a cost of around $4 million.[4] It comprised accommodation for 1000 service personnel, a hospital, an airstrip, power generators and a distillation plant to convert salty bore water to fresh drinking water.[1]

Nuclear-weapons testing commenced at Maralinga on 27 September 1956.[5] Over the next 13 months, seven nuclear explosions were made variously from towers, the ground surface, an air drop and beneath balloons (Table 8).[5] Their explosive yields ranged from 1 to more than 25 kilotons.

One Tree, the first bomb in the Buffalo series, was exploded from a tower at 5 p.m. on 27 September 1956. As part of the test, 283 military officers, known as the 'Indoctrinee Force', were stationed 8 kilometres upwind from the tower at the time of the blast.[4,7] Force members inspected both the explosion area and the effect of the blast on target response items between one and four days after the One Tree firing.[7] Marcoo was exploded on the ground seven days later. To watch the blast, 110 Force members were positioned: one member occupied a Centurian tank, 24 remained in covered shelters 2 kilometres from ground zero, while 82 stood in the open, just less than 3 kilometres from the firing site. On 11 October, Kite was dropped from an RAF Valiant bomber at 9000

Len Beadell

The majority of Army personnel involved in the nuclear testing program at Emu and Maralinga were accommodated in tents.

metres. The bomb was detonated 150 metres above the ground after 45 seconds free-fall.[4] An unexpected wind change resulted in some radioactive fallout settling over the Maralinga village about 10 hours after the explosion.[7] The final bomb in the series, Breakaway, was exploded from a tower five minutes after midnight on 22 October to take advantage of favourable weather conditions.

Eighty-six ground stations were erected around Australia to monitor radioactive fallout during the Buffalo series.[4,7] Each station comprised a metal post and plate, onto which 'sticky paper' was attached to collect deposited fallout. Seventy-six of the stations also had air pumps with filters to determine the amount of radioactivity which would have been inhaled by people at these locations.[7]

TABLE 8: Summary of British Nuclear-Weapons Testing in the Great Victoria Desert[a]

Operation	Code name	Place	Date	Type	Yield (kilotons[b])
Totem	Totem 1	Emu	15.10.53	Steel tower, 30 m	10
	Totem 2	Emu	27.10.53	Steel tower, 30 m	8
Buffalo	One Tree	Maralinga	27.9.56	Aluminium tower	15
	Marcoo	Maralinga	4.10.56	Ground surface	1.5
	Kite	Maralinga	11.10.56	Air burst (free-fall)	3
	Breakaway	Maralinga	22.10.56	Aluminium tower	10
Antler	Tadje	Maralinga	14.9.57	Aluminium tower	1
	Biak	Maralinga	25.9.57	Aluminium tower	6
	Taranaki	Maralinga	9.10.57	Air burst (balloon)	26.6

[a]Adapted from the Technical Assessment Group report.[10]
[b]One kiloton is equal to 1000 tons of conventional explosive, TNT.[1]

Several new committees were established during the 11-month delay between Operations Buffalo and Antler. The Maralinga Safety Committee was formed to give Australia greater control over the conduct of atomic weapons tests at Maralinga, particularly from the aspect of public safety. The National Radiation Advisory Committee became responsible for advising the Federal Government on radiological surveys and measurements in Australia.[4]

Operation Antler finally commenced in September 1957. Tadje, the first bomb in the series, was detonated from a tower on 14 September 1957. Its yield was the lowest recorded in the Maralinga atomic trials (1 kiloton). However, cobalt 60 had been included in the bomb as a tracer. Small radioactive pellets containing this component were found scattered over the ground during a routine survey of the bomb site in early July the following year.[4] Biak was fired from a tower 11 days later, after a delay caused by unfavourable weather conditions. Taranaki was exploded from a balloon suspended 300 metres from the ground on 9 October. The airburst yielded 26.6 kilotons.[7] Taranaki was the final nuclear test carried out in Australia.

A total of 12 nuclear explosions had been conducted at Emu, Maralinga and the Monte Bello Islands between 1952 and 1957.[4,5,7] The British nuclear testing program was judged as a resounding success by its international peers.[1] The United Kingdom, the United States and the former USSR entered into a three-year moratorium on atmospheric nuclear weapons testing in November 1958.[3,4]

The atomic explosions at Emu and Maralinga deposited radioactive fallout around these two locations, and in some instances caused glazing of sand and induced radioactivity in the soil.[6] Small pellets of cobalt 60 were also dispersed at Tadje. Clouds from the series of explosions over the five-year period had spread low-level radioactive material thinly over most of the continent. However, the major legacy of this period of activity in the Great Victoria Desert was the residual plutonium contamination which occurred during so-called 'minor trials'.[4,5,7,9]

'Minor' Trials

Minor trials began at Emu in 1953, resumed at Maralinga in 1955, and continued until 1963. The exact number of minor trials conducted may never be known, although estimates range up to 600.[7] The minor trials were not nuclear explosions, but they did involve the dispersal of some nuclear materials such as plutonium and uranium and other toxic elements including beryllium.[5,7] Most of the minor trials were carried out during the period of international moratorium from November 1958 to September 1961.[1]

The early minor trials were given the code names Kitten, Tims and Rats. They were designed to test individual components or sub-assemblies of nuclear weapons.[7] The later Vixen trials were basically safety assessment tests. Vixen A series was conducted to investigate the dispersal of radioactive and toxic material following an accident. Vixen B was carried out to determine the effects of accidental detonation of some of the high explosives in a nuclear weapon.[4,7]

The Kitten trials did not involve the use of plutonium. Five Kitten experiments were conducted at Emu from 26 September to 17 October 1953.[5] They represent the only minor tests held at this location. Another 94 trials in the Kitten series were later held at Maralinga. The Tims, Rats and Vixen trials were all carried out at Maralinga.

In the 1960–61 Tims tests, 1.2 kilograms of plutonium were used at sites known as TM100 and TM101. A further 0.98 kilograms were used in the Vixen A trials at Wewak between 1959 and 1961.[7] Approximately 22 kilograms of plutonium were dispersed explosively in various plumes as a result of the Vixen B trials conducted at Taranaki in 1960, 1961 and 1963.[7,10] Plutonium was scattered as a finely divided dust, as small sub-millimetre particles, or as surface contamination on larger fragments of debris.[10]

Precisely how much plutonium remains at Maralinga as a result of the Tims and Vixen trials is not known. Twenty-one burial pits at Taranaki are believed to contain 2–20 kilograms of plutonium associated with an estimated 830 tonnes of debris and 1120 tonnes of soil from the Vixen B trials.[10] Four further pits outside Taranaki contain around 90 grams of plutonium.[10] The full extent of plutonium dispersal by wind or dust storms may never be known. Plutonium has a half-life of 240,000 years, that is the time taken for half its radioactivity to decay. It is hazardous to humans if ingested or absorbed into the body, particularly by inhalation of contaminated dust particles.

In addition to plutonium, up to 7 tonnes of uranium are buried in pits at a site known as Kuli.[10] Around 101 kilograms of beryllium are believed to have been used in minor trials, 99 kilograms of which were dispersed across Maralinga.[7] Only 1.8 kilograms of beryllium are known to have been recovered.[7]

Repercussions of the British Nuclear Program

Clean-up and Monitoring

A series of clean-ups at the Maralinga and Emu test sites took place during the 1960s and included Operation Cleanup (1963), Operation Hercules V (August to November 1964) and Operation Brumby (April to July 1967).[7] The latter was carried out by a British team from the Atomic Weapons Research Establishment (AWRE), under guidelines laid

Len Beadell

Britain's chief nuclear scientist throughout the atomic testing program was Sir William Penney (pictured third from the left).

down by the Australian Atomic Weapons Test Safety Committee (AWTSC).[7] Plutonium that could not be recovered or buried at Taranaki was simply ploughed into the topsoil to a depth of only 15–25 cm using a scraper and grader.[7] Contaminated soil at TM100 and TM101 was treated in a similar manner.

The Australian Atomic Weapons Tests Safety Committee (AWTSC) reported to the Prime Minister in July 1967. The Committee stated that Operation Brumby was successful, that the Maralinga and Emu test sites were deemed to be radiologically safe, and that permanent and unrestricted access could be made to all but a few small areas.[7] A descriptive report on Operation Brumby was prepared by a British scientist, Noel Pearce, in 1968.[11] The complete, unedited Pearce Report was not tabled in Australian Parliament until 4 May 1984.[5] The 1984 Royal Commission (see later) concluded that Operation Brumby did not satisfactorily decontaminate the Maralinga Range.[7] There is also considerable current scientific debate as to whether Operation Brumby helped or hindered any future clean-up of Maralinga.[6]

On 23 September 1968, an agreement was signed between British and Australian authorities releasing the United Kingdom from all future liabilities and responsibilities with respect to decontamination of Maralinga.[7]

The Australian Ionising Radiation Advisory Council (AIRAC) undertook a study on radiological safety and future land use at Maralinga in August 1977.[12] Its report was tabled in Federal Parliament during 1979.[5] AIRAC stated, among other things, that levels of radiation at four ground-zero sites were too high for continuous human occupation.[5]

Five hundred grams of plutonium, buried in an underground pit near the Maralinga Airfield cemetery, was returned to the United Kingdom during 1979.[5,7] The plutonium had been used during six Tims trials carried out in 1961.[7] The removal of the plutonium from the burial site took 12 days, during which time oppressive temperatures and strong winds made working conditions difficult.[1] The 500 grams of plutonium, dispersed in 1.63 tonnes of salt and sealed in six steel drums, were flown by Hercules aircraft from Maralinga to Edinburgh Air Base in March 1979.[1,5] It was then transferred to an RAF VC10 aircraft for transport to England.[1]

In September 1980, AIRAC was asked by the Minister for Science and Environment to prepare a major report on the following aspects of the British nuclear program: the nature and distribution of radioactive fallout from the tests, the effectiveness of safety measures for the protection from radiation injury of Australian personnel involved in the tests, and the impact of the tests on the health of the Australian public.[7] The Committee's report was published in January 1983.[7,13]

The Australian Radiation Laboratory (ARL) conducted a radiological survey at Emu and Maralinga between February 1984 and April 1985.[7] Scattered pieces of steel shrapnel contaminated with plutonium oxide were found at Taranaki, TM100, TM101 and Wewak.[5] It was estimated conservatively that there were between 25,000 and 50,000 plutonium-contaminated fragments in the Taranaki area.[7] Contamination from one plume extended north-west from the Taranaki firing pad for at least 18 kilometres.[7] The total amount of plutonium distributed on the surface around Taranaki (including the fragments and the north-west plume) was estimated to be around 1.4 kilograms, suggesting that more than 90 per cent of the 22 kilograms of plutonium used at Taranaki ended up in the burial pits.[7] Fences and warning notices were placed at contaminated sites during July/August 1984.[5]

In May 1984, the Federal Government under Prime Minister Hawke established another committee, headed by Sydney professor, Charles Kerr. Its charter was to urgently review all available data on atmospheric fallout resulting from the British nuclear tests. The Committee was given only 16 days to carry out its work. Kerr's report was critical of the AIRAC's 1983 paper and became the catalyst for the establishment of a Royal Commission.[7,14]

The Human Impact — Aboriginal People

Once the British nuclear testing program was announced, moves intensified to 'block' Aboriginal movement through the Great Victoria Desert. The Australian Department of Supply briefed residents of pastoral properties along the desert's periphery.[5] Station managers agreed to notify the Security Office at the test ranges of all persons arriving or passing through their properties.[5] The practice of ground patrol through the desert was continued by Walter MacDougall during the Emu and Maralinga trials. A second Patrol Officer, Robert Macaulay, was appointed in August 1956 to assist MacDougall in monitoring Aboriginal movements in and around the Great Victoria Desert.[1,7] Macaulay, who was based at Giles Weather Station in the Rawlinson Ranges, was basically responsible for the western half of the desert, while MacDougall patrolled the eastern section. Air patrols were also carried out at Maralinga.

Despite the best intentions of MacDougall and Macaulay, the patrol's task of locating Aboriginal people in such a vast area was like finding a needle in a haystack. A Patrol Officer was not stationed permanently at either Emu or Maralinga during the entire test period.

There were also several distinctive facets of Aboriginal lifestyle that predisposed them to increased risk from radioactive fallout. Aboriginal people were generally barefooted, wore little or no clothing, lived in the open, and gathered food and water from sources which were often exposed to the elements of nature.[1]

The 1984 Kerr Committee concluded that 'the probability of Aboriginal movement through the nuclear ranges during the test period was higher than previously acknowledged'.[14] There are a number of records of Aboriginal people being encountered in the Great Victoria Desert (by Patrol Officers and others) well after the commencement of the nuclear tests and at least up until the early 1960s.[1,5,7,9,15,16] Aboriginal people are now known to have 'come out' of the Great Victoria Desert for the first time as recently as the late 1980s.[17]

Two well-documented incidents that occurred during the nuclear testing program illustrate some of the above points.

The first is the record of an Aboriginal family actually entering the Maralinga firing range, an event which became known as the 'Pom Pom' incident.[5,7] Charlie and Edie Mulpuddie, together with their son Henry and young daughter Rosie, had undertaken a year-long, southerly trek from the Everard Ranges towards Ooldea following traditional waterholes.[3,7] They were unaware of Ooldea's closure and their route took them into the Maralinga area. On the morning of 14 May 1957 an astonished army officer (Captain R. Marqueur) found Charlie near the crater formed by the explosion at Marcoo seven months earlier.[7] Following discussions with the family, it was estimated that they had walked across 1.6 kilometres of contaminated ground.[5] All members of the Mulpuddie family were immediately monitored for radioactivity and underwent decontamination procedures. They were then driven to Yalata. Edie, who was pregnant at the time, gave birth to a stillborn baby at Yalata.[3] Her next child, Allen, was born in 1961 but died of a brain tumour two years later.[1] Edie's daughter, Rose, lost a child herself in 1973.[1]

The so-called 'black mist' incident represents one of the most difficult issues of the British nuclear testing period. 'Black mist' is the name that has become widely used to describe the radioactive cloud discharged from the firing of Totem 1 at Emu on 15 October 1953.[7] The cloud travelled north-east from Emu and passed over Wallatinna and Welbourn Hill (near Marla). It was witnessed by a group of around 30 Aboriginal people who were living at Wallatinna.[5] Several members of the group later gave evidence to the 1984 Royal Commission. The cloud was described variously as 'hazy, like a fog', 'had a strong smell', 'some cloud of some sort, or a mist' and 'black-like smoke…it was quiet, just moving'.[7]

Some days after the black mist had disappeared, people began experiencing sore eyes, skin rashes, coughing, vomiting and diarrhoea.[5] Yami Lester, now a prominent member of the Australian Aboriginal community, was around 12 years old at the time and living with his family at Wallatinna.[7] He stated at the Royal Commission that his sight was temporarily lost in both eyes soon after the mist had passed. Vision was later regained in his left eye, but he remained blind on the right side.[3] He lost his sight completely in 1957.[7]

The origin of the black mist and its consequent effects on Aboriginal health have been the subject of considerable scientific debate. Meteorological studies, together with mathematical and statistical modelling, have indicated that a black mist passing over Wallatinna and Welbourn Hill was possible.[7] The Royal Commission concluded there was no reason to disbelieve Aboriginal accounts that the black mist had occurred and that it made some people sick.[7] The sickness experienced by Aboriginal people may have resulted from radiation exposure, a psychogenic reaction to a frightening experience, or a combination of both.[7] The Commission was unable to decide whether the black mist caused or contributed to the blindness of Yami Lester.[7]

In 1981, the South Australian Health Commission attempted to investigate the effects of the nuclear testing program on Aboriginal health.[18] The study identified 30 cases of cancer among Pitjantjatjara people living at Yalata and the North West Reserve between 1969 and 1980. However, the study was severely hampered by the lack of health records available for the test period, and the Aboriginal population at risk could not be accurately defined.[7] The Royal Commission's final report concluded that there was 'little prospect of carrying out a worthwhile epidemiological study of those [Aboriginal] people involved in the tests nor of others who might have been directly affected by them'.[7]

The Human Cost — Service Personnel

More than 20,000 British service personnel were involved in the United Kingdom's nuclear testing program in Australia and the Pacific region. Approximately 15,000 Australians participated in the tests in the Great Victoria Desert and the Monte Bello Islands.[1]

There can be no doubt that some of these personnel were exposed to radioactive contamination, but the link between exposure and longer-term health problems such as cancer has been difficult to substantiate.

In 1982, Dr John Donovan, from the Commonwealth Department of Health, conducted a health survey among Australia's veterans.[19] The survey was in two parts. A questionnaire was posted to veterans seeking their health histories and possible indicators of exposure to radiation during the tests. Causes of death, based on the death certificates of those who had died, were also analysed.[1] The Donovan report concluded that Australians who participated in the nuclear testing program had not suffered significant adverse health effects.[1,5,19] However, this conclusion was later questioned by the Royal Commission, which regarded the report as an inadequate epidemiological study due to the paucity of relevant information on which it was based.[7] Only 16 per cent of veterans responded to the postal questionnaire.[1]

Results of a follow-up study into the mortality and incidence of cancer among British veterans was published in the *British Medical Journal* during December 1993.[20] The seven-year study was carried out by leading British cancer specialists and members of the United Kingdom's National Radiation Protection Board. Cancer rates among 21,358 ex-servicemen who took part in nuclear tests in Australia and the Pacific were compared with a control group of 22,333 regular service personnel. The study's conclusion was:

> *Participation in nuclear weapon tests had no detectable effect on expectation of life or on subsequent risk of developing cancer or other fatal diseases. The excess of leukaemia in participants compared with controls seems to be principally due to a chance deficit in the controls, but the possibility that participation in the tests may have caused a small risk of leukaemia in the early years afterwards cannot be ruled out.*[20]

Archie Barton (left) and Maralinga Tjarutja elders, Hughie Windlass (centre) and Barka Bryant (right), visited London in 1993 to speak to British Government officials regarding financial arrangements for an appropriate cleanup of the Maralinga test site.

Maralinga Tjarutja

By early 1986, over 200 compensation claims had been lodged by Australian veterans over illness or death allegedly suffered as a result of nuclear testing.[1] Some civil court actions have also been taken. Few claims have been successful.[1]

The human 'cost' of the British nuclear weapons testing program conducted in the Great Victoria Desert will remain one of the most vexed public health issues of modern time.

Royal Commission

A Royal Commission into the British nuclear testing program was announced by the Federal Government under Prime Minister Hawke in July 1984. Its official opening was held in Sydney on 22 August the same year.[7] The Hon. James McClelland, Dr Bill Jonas and Jill Fitch were appointed Commissioners. Counsel assisting the Commission was Peter McClellan. James McIntyre and Robin Auld acted on behalf on the Australian and British Governments, respectively. Aboriginal people were represented by Geoff Eames and Andrew Collett.

The Commission sat for 118 days. Oral evidence was given by 311 witnesses, including scientists, technicians, veterans and Aboriginal people. The broad-ranging enquiry was conducted in Sydney, Melbourne, Adelaide, Brisbane, Perth, Karratha, Maralinga, Marla, Wallatinna and London. There were 918 exhibits and the transcript of the Commission proceedings amounted to 10,424 pages.[7] Two hundred and one conclusions and seven recommendations were made.[1,7]

The Commission decided that the British Government should bear sole financial responsibility for future radiological clean-ups at Maralinga, Emu and the Monte Bello Islands. It stated that action should be commenced immediately to clean up Maralinga and Emu to the satisfaction of the Australian Government, and to ensure that unrestricted habitation by traditional Aboriginal owners could be resumed as soon as possible. The Australian Government was deemed accountable for compensation to those Aboriginal people who had a traditional interest in sites within the former Maralinga Prohibited Area for 'loss and enjoyment of their lands since the beginning, and as a result of the atomic tests program'.[7] The Commission advocated compensation should take the form of technology and services necessary to re-establish traditional links with the land. It also recommended a Maralinga Commission be formed to determine options for an appropriate clean-up, and to oversee the future rehabilitation and management of the test sites.

However, in 1986 the Australian Government formed its own advisory body, known as the Technical Assessment Group. Its tasks were to undertake a series of scientific studies at Maralinga and to assess options for an environmental and radiological clean-up.[6,10] The Group comprised two Australian, two British and one American scientist.[6]

Anthropological, radio-ecological, bioavailability, inhalation hazard, radiochemical and chemical studies were conducted to establish baseline data to determine a level of acceptable contamination for a series of land-use options.[10] An aerial radiological survey provided contamination contours at both Maralinga and Emu.[10] A dosimetric modelling study brought all the data together and assessed the radiation hazards to people occupying the area, including Aboriginal people living in a semi-traditional lifestyle.[6,10] A further set of studies provided background data for the determination of rehabilitation options and their associated costs.[10] The Group's final report was completed in 1990.

It stated that there were still approximately 100 square kilometres of land around Maralinga that were contaminated to a level exceeding 5 millisieverts per year. The report estimated the incidence of fatal cancer in Aboriginal people living a semitraditional lifestyle at Oak Valley Community (north-west of Maralinga) would amount to one to two cases per hundred years from an annual committed dose of 5 millisieverts. (This dose was considered to represent the borderline between acceptable and unacceptable radiation risk for people living such a lifestyle.)[10] It also noted that radiological hazards could still result from the inhalation of contaminated dust particles (particularly by young children) or from the ingestion of contaminated soil (by infants).[10]

The report listed a series of clean-up options and associated costs, ranging from $13 million for fencing and continuous surveillance to $650 million for a total radiological clean-up.[6,10]

The traditional Aboriginal owners of the land, Maralinga Tjarutja (see Chapter 11), considered the alternatives presented in the report and accepted a $100 million clean-up option. The British Government was subsequently lobbied by a series of Australian political figures and Aboriginal representatives including Mr Archie Barton, Maralinga Tjarutja's Administrator. In June 1993, the United Kingdom agreed to contribute £20 million (around $A45 million) to clean up Maralinga.[21–23] The Australian Federal Minister for Primary Industries and Energy, Senator Collins, stated in April, the following year, that a $10-million, five-year contract to clean up Maralinga had been awarded to Australian Construction Services.[24]

Maralinga Tjarutja have also sought compensation from the Australian Government for the risk and disadvantage of having to live adjacent to contaminated land for the next 240,000 years, and for the loss of the use and enjoyment of that land.[6] Senator Collins announced on 2 December 1994 that Maralinga Tjarutja would receive $13.5 million compensation from the Federal Government.[25] This agreement ended a decade of long, but peaceful, negotiation between the Maralinga people and the Australian Federal Government.

The last phase of this controversial period of the Great Victoria Desert's history will not be closed until the Maralinga clean-up is finally completed.

Mark Shephard

George Wood

Len Beadell

TOP: Len Beadell signing the visitors book at Neale Junction in the heart of the Great Victoria Desert in May 1988.

ABOVE: In 1951, Beadell positioned a series of observation posts at set distances along the centreline of fire from Woomera Rocket Range. This picture shows the 400-mile post in the north-western part of South Australia. White canvas sheets, seen in the background, formed a cross that was visible from the air.

LEFT: Beadell nailed an aluminium plate to each observation post. The plate shown details the exact latitude and longitude of the 250-mile post.

Chapter 10

THE BEADELL TRACKS

Len Beadell was a truly remarkable Australian. People who knew Len respected him immensely as a man who never spoke unkindly of anyone and who had an unfailing sense of optimism, a quick wit and an infectious laugh. He was most at home 'in the bush', where he felt comfortable in his trademark attire of well-worn hob-nailed boots, no socks, and loose-fitting shorts. Len spent 10 years of his life building a network of roads which opened up 2.5 million square kilometres of the Great Victoria, Great Sandy and Gibson Deserts. During this period, he worked as Range Reconnaissance Officer for the Department of Defence (formerly the Department of Supply). He received a British Empire Medal in 1958 and the Order of Australia in 1988 in recognition of his and his team's now legendary road-making feats. The story of Len's work has been vividly and humorously detailed in six best-selling books which he authored.[1–8] *However, a book on the Great Victoria Desert would be incomplete without a short account of Len Beadell's invaluable contribution to the area.*

At the end of World War II, the British and Australian Governments agreed to establish a rocket range on mainland Australia. A large amount of initial survey work was required to find a suitable site for the range. Colonel Fitzgerald, Director of the Australian Army Survey Corps, approached Beadell to see if he would be interested in the survey project. Beadell was well credentialled for the position. He had served in the Army Survey Corps in New Guinea near the end of the war and had recently returned from a CSIRO scientific expedition to Arnhem Land. Beadell was appointed to the task, following an interview with General Sir John Evetts from the British Army.

Beadell selected sites for the township of Woomera and a rocket launching pad during 1947.[9–10] A flight path for rockets to be fired from Woomera was then chosen. It extended north-west from Woomera, across the Great Victoria, Gibson and Great Sandy Deserts, to a point on the 90 Mile Beach between Port Hedland and Broome in Western Australia. It was initially intended that tracking instrumentation would be placed along the centreline of rocket fire, at predetermined distances from the Woomera launching pad. Beadell was given the task of leading field trips to locate and mark these positions, or 'observation posts' as they became known.

In 1951, Beadell and a small party of scientists left Mabel Creek Station (50 kilometres west of Coober Pedy) and headed north-west into the Great Victoria Desert. Their aim was to reach the first observation post, at the 250-mile point from the launching pad. Cloudy weather hampered Beadell's astronomical observations, but he eventually guided his party to the location. A thick mulga post, marked with the date and year, was placed in the ground. Sheets of white canvas were laid for a distance of 30 metres from the post to form a cross which would be visible from the air. Beadell later positioned

Len Beadell

ABOVE: Beadell refound Tallaringa Well during 1951. The well is now the most widely known landmark in the Tallaringa Conservation Park.

RIGHT: Beadell and his party dug Tallaringa Well to reach water and then timbered its sides.

Len Beadell

similar posts at the 300-, 400- and 500-mile points, the latter being close to the north-west corner of South Australia. (All the observation posts remain standing today.) Mosquito bombers were now able to fly along the centreline following and photographing the crosses. The aerial photographs were to be used to construct a strip map and plan tracking stations. Despite Beadell's efforts to establish these field markers, they were never used for their intended purpose, as more technologically advanced equipment was soon developed at Woomera. Nonetheless, this period represented Beadell's introduction to the Great Victoria Desert.

Beadell also made a concerted effort to relocate Tallaringa Well on the eastern fringe of the desert during 1951. He knew of the well's existence by studying Richard Maurice's journals, and its position was tentatively marked on an old aeronautical map. By taking repeated star observations with his trusted theodolite, Beadell finally pinpointed its position on the ground. The well was a small 'dimple', 1 metre deep and 3 metres wide. It was silted over and badly polluted with the decayed remains of kangaroos and dingoes. His party dug the well to a depth of 4 metres, when water was struck.

In 1952, Beadell was summoned to the headquarters of the Long Range Weapons Establishment at Salisbury in South Australia. He was told under a veil of secrecy that 'it had been decided to detonate an atomic bomb in Australia'.[2] Beadell was allocated the task of selecting a suitable site for the blast. His only guidelines were that the site should not be closer than 30 kilometres at right angles to the centreline of the missile range and that there should be sufficient flat area around the site to establish a base and land aircraft.

A six-month search of country in the north-eastern Great Victoria Desert brought Beadell to a long claypan, which he named Dingo. It was large enough to land planes, but was only 1 kilometre from the centreline and 10 kilometres from the 300-mile point. Beadell used Dingo Claypan as a base from which he then radiated up to 120 kilometres in all directions looking for a better site.

A large open area devoid of sandhills was encountered around 30 kilometres in a south-south-west direction. It was perfect for an explosion tower. Within 10 kilometres of this open area there was a hard claypan nearly 2 kilometres long by 1 kilometre wide, on which aircraft of any weight could land. A small rise overlooking the claypan provided an ideal location for a village and administration centre. While camped in the area, Beadell befriended an emu that emerged daily from the vegetation around the claypan and fed on scraps. The site now had a name — Emu.

Work at Emu began in earnest. A 60-metre bore was sunk, but its water supply was so salty that a distillation plant had to be erected. Emu village sprang up and a road was made between the claypan and the forward area. Beadell had a major input into a regular newsletter, known as *Bulldust*, which was produced for the 400 service personnel stationed at Emu.

In 1953, Beadell constructed a road from Mabel Creek Station (50 km west of Coober Pedy) to Emu. The crossing of Mabel Creek proved difficult because of recent heavy rainfall in the desert.

Len Beadell

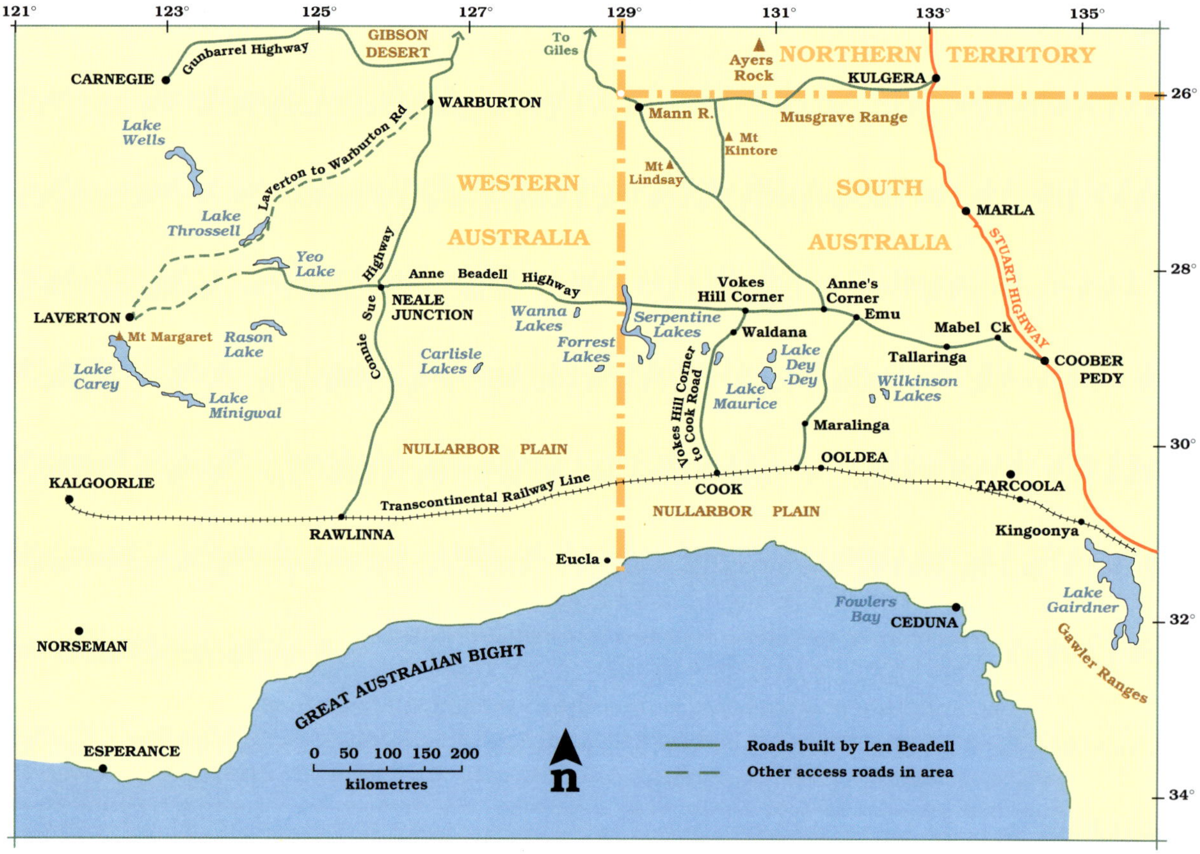

The network of roads built by Len Beadell in and around the Great Victoria Desert.

Ground access to the Emu area was urgently needed, and Beadell spent the early part of 1953 constructing a road from Mabel Creek, past Tallaringa Well, to Emu. This small section of road was the beginning of the Anne Beadell Highway, which would later link Coober Pedy with the Western Australian goldfields.

As described previously, two atomic bombs were detonated at Emu in October 1953. Beadell stood next to Sir William Penney, Britain's chief nuclear scientist, at the time of the Totem 1 explosion. He vividly recalled the feeling of intense heat at the back of his neck immediately after the bomb's detonation and the loud boom from its shock wave 20 seconds later. For Beadell, it was the climax of two years' solid work and the most exciting moment of his life.

The site for future atomic testing at Maralinga was chosen during the Totem 1 series. Beadell's search for what eventually became Maralinga took him on an 800-kilometre cross-country reconnaissance past Lakes Maurice and Dey-Dey, south to the Nullarbor Plain, along the Transcontinental Railway Line and north to Tietkins Plain. He visited Ooldea Soak during this search. The deserted mission buildings were now almost two-thirds engulfed by sand.

Much of Beadell's time during 1954 was spent preparing a contour map and developing the site of Maralinga. Heavy equipment remaining at Emu was needed at

Maralinga, and in August Beadell became responsible for building a road to link the two sites. During this work he found and named Observatory Hill, a 'bowl-shaped rocky knob' south of Emu.[3] He also stumbled upon an Aboriginal ceremonial ground, which he later called 'Aboriginal Stonehenge'.[3] (Today this feature cannot be visited, unless prior written permission is obtained from Maralinga Tjarutja — see Appendix.)

During Beadell's stay at Maralinga, he assembled a party of six men to assist with future road-building projects. Doug Stoneham and Scotty Boord drove the bulldozer and grader, respectively. Rex Flatman was the team's mechanic. Bill Lloyd and Frank Quinn were long-distance supply drivers, responsible for providing the crew with fuel, water and food. Paul Christensen acted as cook. This party became affectionately known as the Gunbarrel Road Construction Party and most of its members stayed with Beadell during 6000 kilometres of subsequent road-making across Australia's Central and Western Deserts. In building these roads, Beadell would initially undertake a solo reconnaissance of up to several hundred kilometres to survey the country ahead and establish the most suitable course for a road. He would then return to the construction party's base and lead the road-making team through the surveyed area. A flashing mirror was used to direct the party in thin scrub, and a pistol flare was fired if thick vegetation was encountered.

A weather station was needed in 1955 to assist scientists at Maralinga to determine favourable atmospheric conditions for the detonation of bombs. Beadell's party bulldozed a road from Victory Downs just over the Northern Territory–South Australia

Beadell named Anne's Corner, west of Emu, after his wife Anne. Anne Beadell (left) is pictured at the Corner with the author's wife, Anne, and son, Matthew.

Mark Shephard

border, west through the Musgrave, Mann and Tomkinson Ranges to the south-eastern edge of the Rawlinson Ranges, where Giles Weather Station was established. This road formed the initial part of Beadell's most well-known landmark, the Gunbarrel Highway. The Gunbarrel Road Construction Party's grader is now permanently housed at the Giles Weather Station.

Further roads were also required in the Great Victoria Desert. Scientists wanted to capture radiation fallout north and west of Maralinga using monitoring stations described previously as metal posts and plates onto which 'sticky paper' was attached (see Chapter 9). Beadell therefore constructed a road west from Emu for a distance of 110 kilometres; this, in effect, became the first extension of the Anne Beadell Highway into the desert's heart.

In 1957, work commenced on another road from Mt Davies, in the Tomkinson Ranges, south-east across the northern Great Victoria Desert. It intersected the new extension of the Emu road around 50 kilometres west of the former bomb site, at a point now known as Anne's Corner. Mobile meteorological stations needed to be established along the route and Geiger-counter readings were required after each bomb test at Maralinga. Beadell's initial reconnaissance along this 350-kilometre stretch of country between Mt Davies and Emu was particularly arduous. Major mechanical and fuel problems were encountered and a barrage of 200 or more towering sand dunes were met in a 130-kilometre section of the Great Victoria Desert. The desert country along Beadell's path was nonetheless breathtaking. Mt Lindsay, a prominent hill in the Birksgate Range, provided him with a spectacular panorama of the northern fringe of the Great Victoria Desert. Beadell's vehicle finally succumbed to a broken rear axle in the midst of desert sandhills some 150 kilometres north of the Emu road. He radioed his position to Maralinga and remained at the spot for two weeks until his retrieval.

After a well-earned rest, Beadell returned via the Stuart and Gunbarrel Highways to Mt Davies. He then led the Gunbarrel Road Construction Party down the surveyed line to Anne's Corner. The graded road between the two locations was completed in 1959, after 'diversions' to build an airstrip at Giles and finish the Gunbarrel Highway.

In 1961, Beadell pushed the Emu road further west into the Great Victoria Desert, heading towards an isolated pinpoint on the map named Vokes Hill. He used a series of sun and star observations to slowly draw closer to its plotted position. The base of the 'hill' was reached some 160 kilometres west of Emu. It was not a rocky outcrop, but rather a huge mound of sand with scattered rocks near the top. Beadell scrambled to its summit, where he had an unbroken view to the skyline in all directions. Beadell's mission now was to complete a second road across Australia which would bisect the Great Victoria Desert, approximately half-way between the Nullarbor Plain and the Gunbarrel Highway. A secondary aim was to provide access for parties from the National Mapping Council of Australia, who were endeavouring to prepare accurate survey maps of this hitherto, largely unexplored section of Australia.

Around 10 kilometres south-east of Vokes Hill, Beadell decided on a position for a major road intersection, which he named Vokes Hill Corner (nowadays, it is also widely referred to as Vokes Hill Junction). Construction of the road further west was temporarily abandoned as he now headed due south on a quick reconnaissance to the railway siding of Cook on the Nullarbor Plain. After reaching Cook, Beadell drove directly to Adelaide for an important engagement — his marriage to Anne Matthews!

Len Beadell

Beadell's solo cross-country travel would invariably result in the buildup of twigs and other debris on the windscreen of his trusted Land Rover.

Until 1960, Beadell's Land Rover was his home for 10 months of the year. Holidays were spent visiting relatives and friends or sleeping in his vehicle at the Weapons Research Establishment's headquarters at Salisbury. He decided to buy a house during 1960 and arranged for an agent to let it. He met Anne in mid-1960 while her English family were staying at his home as its first and last tenants. Anne and Len were married one year after their first meeting in the kitchen of his home, where he showed slides on the wall and talked to Anne for hours about his work. He was 38 and she was 20. Anne thought Len was the 'most interesting and exciting man she had ever met'.

Beadell soon returned via Maralinga to Vokes Hill Corner, where construction of the road south commenced. The Vokes Hill Corner to Cook road was needed to provide further access to the Transcontinental Railway Line and to enable Native Patrol Officers to complete a rectangular loop as part of their efforts to monitor Aboriginal movements in the Great Victoria Desert. Road building was temporarily interrupted when Beadell was called to Maralinga and then flown to Adelaide for a surprise visit to his wife Anne. The couple decided that Anne should join Beadell on a future desert excursion. Beadell soon resumed his work on the Vokes Hill Corner to Cook road, and deliberately positioned its course past a series of Aboriginal wells and rockholes, namely Waldana, Churina, Bringyna and Muckera.

Waldana Well was badly silted over and had been difficult to find. Beadell nailed an aluminium plate onto a timber post to mark the site. The plate contained details of Waldana's latitude and longitude, the distance back to Vokes Hill Corner and the date. These aluminium-plated signposts became Beadell's trademark or 'bush signature' and were erected at numerous locations throughout the Great Victoria, Gibson and Great Sandy Deserts. Today they are recognised as major points of interest for desert travellers. Bringyna was found in a small clearing among a horseshoe pattern of gigantic sandhills. These sandhills made access to the well difficult. Beadell therefore positioned the main road around 1 kilometre to the west and cut a small side road to the well. The open, treeless Nullarbor Plain was finally reached and the road to Cook was completed late in 1961. Beadell again drove to Adelaide, this time to be greeted by a small bundle of joy, a daughter Connie Sue.

Beadell returned to Vokes Hill Corner in 1962 to begin the next section of the road west. An initial 160-kilometre reconnaissance brought him to the eastern edge of the Serpentine Lakes. This chain of glistening white saltlakes, which wind for over 100 kilometres in a north–south direction along the South Australian–Western Australian border, presented a potentially formidable obstruction to the road's further progress. At the point where Beadell met the lakes, there was a narrow 100-metre strip of discoloured ground between two large salt pans. Beadell gingerly walked the strip, but was unconvinced it could support the weight of a 15-ton bulldozer. A northerly course along the eastern fringe of the lakes was then taken. After 30 kilometres, he reached a raised stony section of land between two salt pans, which enabled him to drive to the western bank of the lakes. A path across the Serpentines had been found!

Beadell then decided to proceed in a south-easterly direction from the lakes to cut his earlier westerly wheel tracks. This oblique cross-country route across the Great Victoria Desert proved one of the hardest traverses he had ever attempted. He was confronted by 'mountains' of merging sandhills, intense heat, constant thirst, flat tyres and a seized engine. This section of the desert was simply too difficult and uncompromising for future road construction.

Beadell eventually arrived back at the Construction Party's base camp. He led the team forward on his former, westerly course and again met the strip of discoloured land between the lakes. He decided to attempt a crossing at this point to avoid the 30-kilometre northerly detour to the stony land bridge. To the relief and jubilation of all concerned, the bulldozer cautiously 'walked' across the thin neck of land. The South Australian–Western Australian border (longitude 129°E) was within a kilometre of the western edge of the lakes. A suitably inscribed aluminium-plated sign was placed on top of an empty diesel drum to mark the border's position. A bulldozer cut was also made at right angles to the direction of the road to enable future aerial photographs of the border to be taken. The Anne Beadell Highway was now half-completed.

Beadell conducted further reconnaissance west to bypass another saltlake system, the Wanna Lakes. After reaching a point nearly 50 kilometres west of the border, he left the Construction Party and returned via Emu and the Stuart Highway to Adelaide.

Beadell, accompanied by Anne and five-month-old Connie Sue, now headed along the Stuart and Gunbarrel Highways to the Warburton Ranges Mission in Western Australia. Connie sat contentedly in a padded tea chest next to Len and Anne. The trio left the mission and proceeded due south into the north-western Great Victoria Desert. Beadell's intention was to reach a point just over 300 kilometres south, on the same latitude (28°33'S) as the Construction Party. From this point, a 150-kilometre easterly course would lead them to the awaiting team. A week of solid travel brought the Beadells to the southernmost destination, which later became known as Neale Junction. They then progressed east for about 60 kilometres before the vehicle ran out of petrol. The Construction Party moved westward in response to numerous radio messages and pistol flares, and finally rendezvoused with the stranded Beadell family. The entire convoy retraced Beadell's steps to Neale Junction. Neale was the name given to a map of the area

Mark Shephard

ABOVE: A thin strip of land between two saltpans now functions as the main road across the Serpentine Lakes.

RIGHT: Beadell and the author are pictured at the metal drum and aluminium plate that marks the South Australian–Western Australian border.

Anne Shephard

that was drawn around this time by a survey party from Perth. Neale Junction was soon to become a major road intersection.

Len, Anne and Connie Beadell pushed further west from Neale Junction on a reconnaissance to Yeo Lake, where they met a sandalwood cutter's track which led south-west to Laverton. This Sandalwood Road, as it was known, had been made several years earlier by a man named Martin Cable. The Beadells enjoyed the local hospitality at Laverton and nearby Leonora, and soon headed back into the Great Victoria Desert to rejoin the Construction Party. A road was now constructed from Neale Junction to

Warburton, a distance of just over 300 kilometres. The party's heavy machinery was then walked back to the Junction.

It was October 1962. Beadell set his sights on constructing another section of road from Neale Junction south to the railway siding of Rawlinna. The weather at this time of the year was severe, with intense heat and furious dust and hail storms making life uncomfortable. Nonetheless, the task was successfully accomplished. A direct north–south road, which crossed around 650 kilometres of the Great Victoria Desert and linked Rawlinna to Warburton, was now complete. It was christened the Connie Sue Highway.

The final section of road between Neale Junction and Yeo Lake was then bulldozed and graded. The Anne Beadell Highway was completed on 17 November 1962. Construction of this major east–west road link, from Mabel Creek, via Emu, Vokes Hill, the Serpentine Lakes and Neale Junction to near Yeo Lake, had spanned almost 10 years. Both the Connie Sue and Anne Beadell Highways were now 'open' for thousands of subsequent desert travellers to enjoy. (Today, a number of permits are required to traverse sections of these roads — see Appendix.)

Len Beadell retired from the Weapons Research Establishment in April 1988. The Beadells had two other children, Jackie and Gary, and they too had road features of outback Australia named after them (Jackie Junction and the Gary Highway). Following his retirement, Len continued to visit the Great Victoria Desert as guest narrator with outback safari companies run by Dick Lang, Russell Guest and Peter Vernon. Family trips were also made regularly to retrace and relive memories of Len's road-making days. In 1992 Len's grandson, Mitchell, was christened at Tjintirrgarra (or Djindugara) waterhole near the Serpentine Lakes.[11] The family tradition continues...

Postscript: Len Beadell sadly passed away during the final stages of this book's production. However, Len's spirit will remain forever in the desert.

Chapter 11

Current Aboriginal Land Tenure and a Return to Traditional Lands

The Political Winds of Change

Australian Government philosophy towards Aboriginal people has undergone a series of changes during this century. As described in Chapter 8, early policies focussed chronologically on protection, welfare provision and assimilation.[1] In 1972, the newly elected Whitlam Labor Federal Government introduced a new policy, that of self-determination. Its aims were to provide Aboriginal people with a measure of independence from white Australians and to allow them to determine their own future and style of living.[1] The Government implemented the policy by initiating a series of actions which included legislation to enable Aboriginal communities to incorporate and so manage their own affairs. Incorporated communities were able to elect their own council, adopt an Aboriginal name, and have an adviser rather than a manager or superintendent. All members of the community could be involved in decision-making processes.[1] The Whitlam Government was replaced controversially by the Fraser Liberal–National Country Party Coalition in November 1975. The attitude of the new Federal Government towards Aboriginal people remained similar to its predecessor, although the term 'self-determination' was replaced with 'self-management'.[1]

Aboriginal communities in and around the Great Victoria Desert now had the legislative backing to begin the process of claiming tenure over traditional land and to initiate moves to resume living on that land.

Land Tenure

A large proportion of the Great Victoria Desert is currently under the tenure of several major land-holding Aboriginal bodies, through land rights legislation (in South Australia) and a series of leased Aboriginal reserves (in Western Australia).

South Australian Aboriginal Land Rights Legislation

In South Australia, Anangu Pitjantjatjara was granted inalienable, freehold title to the Pitjantjatjara Lands in the north-west of the State under the *Pitjantjatjara Lands Rights Act 1981*. These Lands encompass a large proportion of the northern Great Victoria Desert in South Australia, as well as the Everard, Musgrave, Mann, Tomkinson and Birksgate

Ranges. Maralinga Tjarutja was granted similar title to the Maralinga Lands in the southern Great Victoria Desert under the *Maralinga Land Rights Act 1984*.

Aboriginal land rights have become a major issue on the Australian political, economic and social agenda. As outlined in Chapter 6, Aboriginal culture is based on a set of principles different from those of European society. Aboriginal 'Law' dictates that its people are the permanent custodians of the land. Aboriginal people have strong moral and spiritual obligations to conserve the land's landscape features, plants and animals, and to perpetuate traditional knowledge of important sites and Dreaming tracks through ceremony and song. Their perception of 'one law, one land, one people' provides the foundation for Aboriginal land rights.[2]

The Evolution of Land Rights

Pitjantjatjara Lands

As mentioned in Chapter 8, nearly 5.7 million hectares of land in the north-west of South Australia (including the northern Great Victoria Desert) were declared the North West Aboriginal Reserve in 1921. Aboriginal people living on the Reserve at Ernabella Mission became increasingly unsettled during the 1960s and began to express the need to be closer to their traditional 'country'.[2,3]

As part of its new policy of self-determination, the Federal Labor Government initiated a Land Rights Commission in 1972 to investigate the issue of land rights for Aboriginal people in the Northern Territory.[1,4] The *Aboriginal Land Rights (Northern Territory) Act* was subsequently passed in 1976.

This legislation provided the impetus for the formation and incorporation of the Pitjantjatjara Council, following a meeting of several communities at Amata in July 1976. The Council resolved to strive for separate South Australian legislation that would

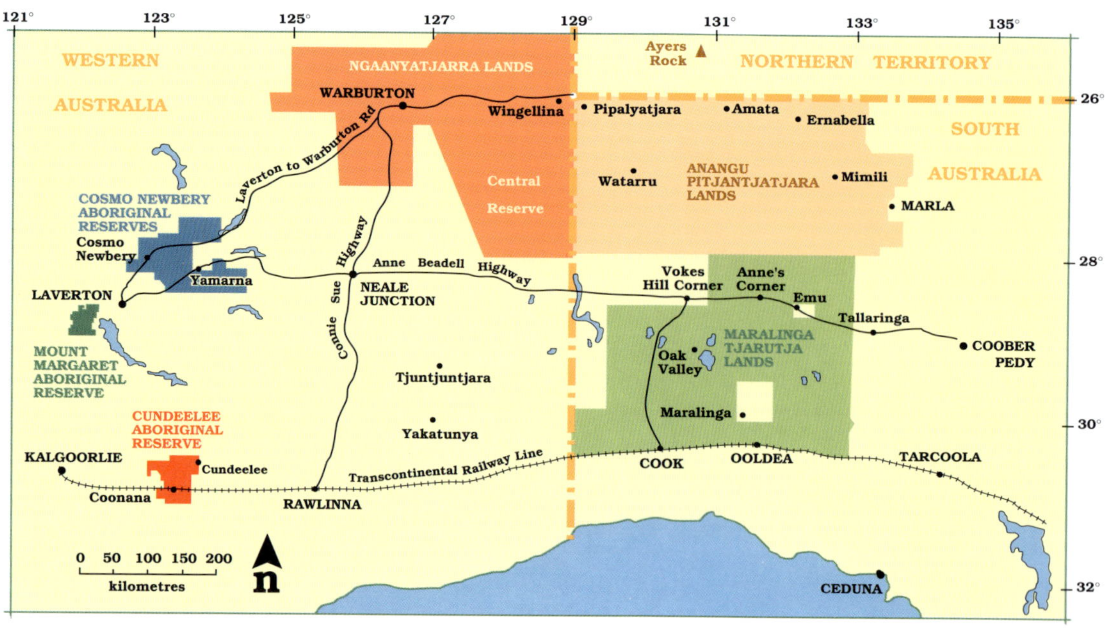

Aboriginal Land in the Great Victoria Desert and the location of selected homelands.

provide their people with perpetual title to 10.26 million hectares of land encompassing the North West Aboriginal Reserve and several adjacent pastoral properties.

The South Australian Premier Don Dunstan set up the Pitjantjatjara Lands Rights Working Party, under the direction of Chris Cocks, on 24 March 1977. The Working Party's report was completed in June 1978.[5] It recommended that new land rights legislation covering the Pitjantjatjara Lands should be enacted and a land-owning body known as Anangu Pitjantjatjara (literally, the Pitjantjatjara peoples) should be formed.[4] This body should be granted inalienable, freehold title to the land and have special controls over access and mining.

Passage of the legislation was delayed for nearly two and a half years and was complicated by a change of State government. The delay precipitated a mass rally of around 130 Pitjantjatjara people and their supporters at the Victoria Park racecourse in Adelaide during February 1980.[2,4] The Pitjantjatjara Land Rights Bill was finally ratified by the South Australian Parliament in March 1981. Premier Tonkin handed title of the land to Anangu Pitjantjatjara on 4 November 1981 at a ceremony held near Ernabella which was attended by 1500 people.[4]

Maralinga Lands

South Australian Premier Sir Thomas Playford gave a commitment to Yalata's Aboriginal people in 1962 that land around the Maralinga area would be returned to the control of its traditional owners after cessation of the British nuclear testing program.[4]

The Commonwealth advised the South Australian Government in 1972 that the Maralinga village was no longer required and that the nuclear-testing area would be de-restricted, except for Section 400 which would be retained by the Commonwealth under a land grant. Section 400 included sites where nuclear explosions had been detonated and radioactive waste had been buried. Later that year, Premier Dunstan offered the Aboriginal Lands Trust freehold title to the entire southern Great Victoria Desert, including the Maralinga area and the Unnamed Conservation Park which had been dedicated two years earlier.[2] Title was to be granted to the Lands Trust on condition that the Yalata Community would receive perpetual lease from the Trust.

However, the land transfer did not proceed as protracted negotiations ensued over residual contamination at Maralinga. A State Liberal Government was elected to office in the interim. It offered the Lands Trust and Yalata Community title over a reduced land package, in which the Unnamed Conservation Park and the eastern Maralinga area were deleted. The Yalata Community received subsequent advice from the Pitjantjatjara Council and the Aboriginal Legal Rights Movement. Lengthy discussions over land title were held at Yalata and Ooldea during 1982.

The Community decided that they wanted separate and distinct title to the 'Maralinga Lands', under the same terms and conditions as the *Pitjantjatjara Lands Rights Act*. An incorporated land-holding body, known as Maralinga Tjarutja, was formed at this time.

The State Liberal Government was defeated at the November 1982 election and Labor returned to power. The Maralinga Tjarutja Land Rights Bill was introduced to State Parliament in May 1983. The new government lacked the numbers to push the bill through the Legislative Council and it was referred to a Select Committee. An amended bill, which granted Maralinga Tjarutja inalienable freehold title to 7.64 million hectares of land, was finally passed on 21 March 1984. Land title was formally

Maralinga Tjarutja

Maralinga Tjarut

ABOVE: Aboriginal people from Maralinga Tjarutja gathered at Ooldea in 1991 for the handover of title to 360,000 hectares of land, which included Ooldea Soak.

RIGHT: Members of Maralinga Tjarutja, namely Peter Pepper (left), Herbert Queama (centre) and Ted Roberts (right), are pictured proudly holding the title documents to this section of their Land.

handed to Maralinga Tjarutja by Premier Bannon at a ceremony near Maralinga on 18 December 1984.

The southern boundary of the Maralinga Lands was drawn along a line 8 kilometres north of the Transcontinental Railway Line. It excluded Ooldea Soak, an area of critical traditional importance to Aboriginal people. An amendment to the original Land Rights Bill was subsequently passed to redefine the southern boundary and to transfer an additional 360,000 hectares of land (including Ooldea Soak) to Maralinga Tjarutja.

While Maralinga Tjarutja was granted inalienable freehold title to the land under the *Maralinga Tjarutja Land Rights Act*, its powers relating to access and mining were less stringent than those conferred under the *Pitjantjatjara Lands Rights Act* (see Chapter 12).

Western Australian Aboriginal Reserves

THE NGAANYATJARRA LANDS

In Western Australia, the Ngaanyatjarra Council (Aboriginal Corporation) was formed in 1981 to represent the land-tenure aspirations of Aboriginal people living in the State's Central Reserves area (which includes a large proportion of the northern Great Victoria Desert).[6] The Council also provides many other functions, including professional and administrative support for its members and the coordination of a range of services (for example road and air transport, purchase of food, building and health programs, and negotiations with mining companies).

The Ngaanyatjarra Council currently comprises 10 communities. Reserve land occupied by each of these communities is leased and managed by the Ngaanyatjarra Land Council (Aboriginal Corporation) via a series of 99-year leases for A Class Reserves (from the Minister for Aboriginal Affairs and the Aboriginal Lands Trust), and via 50-year Special Purpose leases from the Minister for Lands.[6] The Ngaanyatjarra Council, on behalf of and in conjunction with the Ngaanyatjarra Land Council, carries out the Land Council's responsibilities.[6]

Several of the communities have strong connections with the Great Victoria Desert. The Warburton Aboriginal Reserve (277,000 hectares) lies to the north of the Great Victoria Desert, but more than one-quarter of the people living on the reserve have traditional affiliations with the Great Victoria Desert.[7] The large Central Reserve includes a vast section of the northern Great Victoria Desert east of the Connie Sue Highway and north of the Anne Beadell Highway. The 99-year lease to both reserves is held by the Ngaanyatjarra Land Council.[8] The 174,000-hectare Yapupaara Reserve was also recently leased to the Land Council. It is situated on the northern edge of the desert and lies in part across the Connie Sue Highway. Once permanently occupied, Yapupaara Community has been uninhabited for several years in accordance with mourning taboos.[7] However, regular visits by former residents have continued from their temporary homes in Warburton, Coonana and Wingellina.[7] Plans are now being made for the permanent re-occupation of Yapupaara. The area surrounding Yapupaara and land to the north and west of Warburton is currently held under a 50-year Special Purpose lease.

In total, the Ngaanyatjarra Land Council manages 13.1 million hectares of A Class Aboriginal Reserves under 99-year leases and 5.4 million hectares of land under Special Purpose leases.[6] In addition, there is a considerable amount of vacant Crown Land within the Great Victoria Desert (for example west of longitude 125°E and north of latitude 28°S) over which the Ngaanyatjarra Land Council has traditional responsibility.[6]

Nine of the 10 Ngaanyatjarra communities also lie wholly or partly within the Shire of Ngaanyatjarraku local government area.[6] This shire, formerly known as the Eastern Region of the Wiluna Shire, was created as a direct result of a strong commitment by Ngaanyatjarra Council members to gain access to mainstream local government services and programs.[6]

Other Aboriginal Reserves Associated with the Great Victoria Desert

The area around and including the former mission at Cosmo Newbery now comprises four separate C Class Aboriginal Reserves.[8] All the reserves surround the Yamarna pastoral lease, and two abut the Yeo Lake Nature Reserve. The degree of tenure guarantee of a C Class Reserve is less than that of an A Class Reserve, which requires consent of both Houses of State Parliament to change its status.

Mt Margaret Aboriginal Reserve was proclaimed in June 1973 and leased to the Aboriginal Movement for Outback Survival until 30 June 2079.[8,9]

Cundeelee Aboriginal Reserve was also proclaimed in 1973 and leased to Upurl Upurlila Ngurratja Incorporated until 31 October 2085.[8,9]

A Return to Traditional Lands

Under the Federal Government's assimilation policy during the mission era, the desert's Aboriginal people had little autonomy, were alienated from their traditional land, and lived on country that was not their own.[10] The changing government philosophy of the 1970s resulted in a strong desire among many Aboriginal people not only to pursue issues of land tenure, but also to return to their homelands.

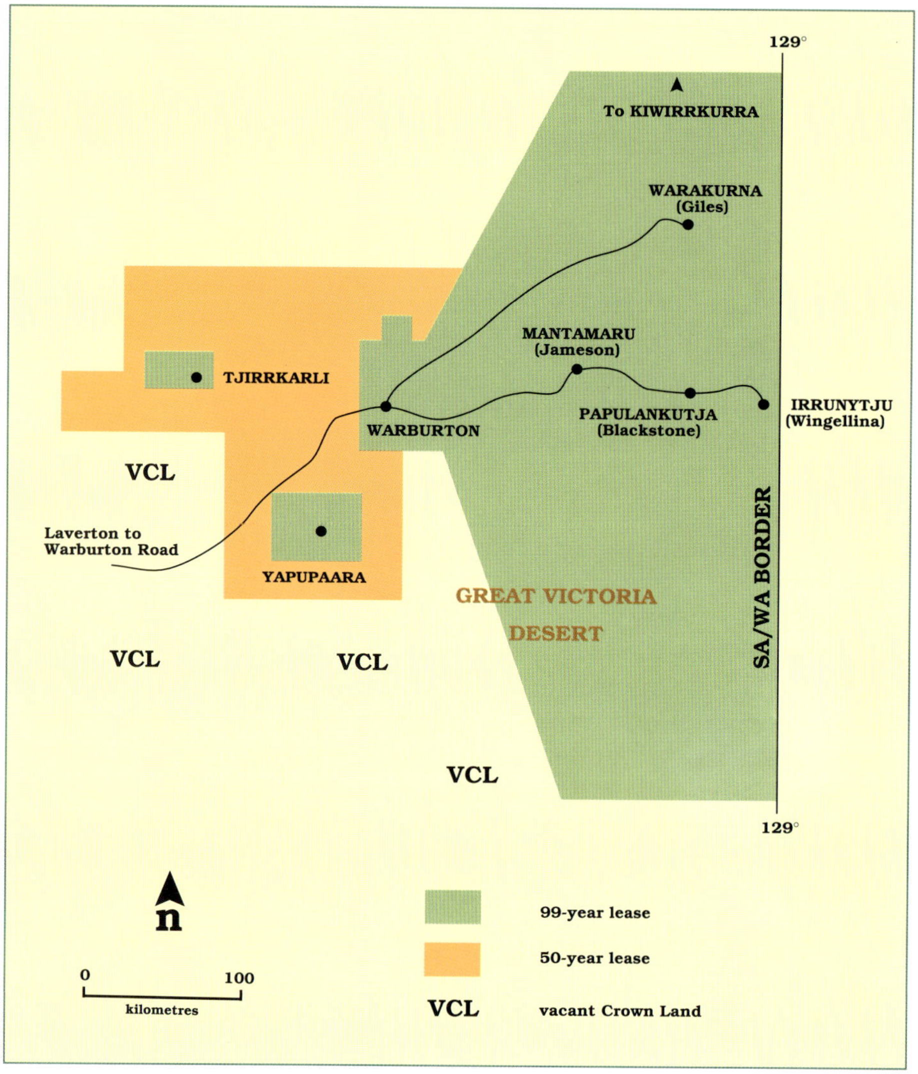

A closer examination of the Ngaanyatjarra Lands.

The homeland or outstation movement among Aboriginal Australians has been described as 'one of the most interesting and dynamic social phenomena in the world'.[11]

The goals of the outstation movement include the stimulation or revitalisation of traditional Aboriginal culture and the removal of social evils associated with white settlement, such as alcohol and petrol-sniffing. These aspirations can only be realised provided that: (i) administrative and organisational changes are put in place to ensure Aboriginal people have real control over community affairs and are able to select their own priorities; and (ii) Aboriginal people have the desire, commitment and confidence to practice self-determination.[10]

The following two examples of homeland establishment in the Great Victoria Desert illustrate both the dedication of many Aboriginal people to this cause and some of the practical problems associated with a return to traditional lands.

Ken Scott

ABOVE: The Ngaanyatjarra Lands encompass a large section of the northern Great Victoria Desert. The northern edge of the desert can be seen from the top of Mt Hinkley, south of Irrunytju (Wingellina).

ABOVE RIGHT: Irrunytju elder Raymond Nelson, with his grandson Anthony. The Irrunytju Community believe that the transfer of traditional knowledge from its elder members to its young children is critically important to the Community's future. Members of the Community regularly visit traditional sites in the Great Victoria Desert.

Anne Scott

Oak Valley

In 1966, the Superintendent of Yalata mission initiated an annual 'bush trip' whereby its Aboriginal residents could visit the Great Victoria Desert for a one- to two-week period.[1] These trips provided the Yalata people with an opportunity to re-establish physical, spiritual and emotional links with their land.[1] The desire to return to 'their country' was seeded during these trips. Aboriginal elders, in particular, wanted to teach traditional Law to Yalata's younger people, who had grown up in a 'foreign' environment and knew little of desert life.

A road was constructed into the Lake Dey-Dey area during the early 1980s by Aquitaine, a French mining company searching for uranium. A group of 66 Aboriginal people determined to leave Yalata in May 1982. They travelled along the Aquitaine road and established a spartan outstation camp at a location the people called Oak Valley (approximately half-way between Maralinga and the Unnamed Conservation Park).[1] Practical difficulties in providing and maintaining basic services such as food, water and medical care made initial life at Oak Valley very tenuous. As a result, their camp was moved several times over the next two years to Pidinga, Ooldea and north of Watson.[12] Following the granting of title to the Maralinga Lands, the party made a firm commitment to establish a permanent outstation or homeland at Oak Valley during March 1985.[1] Today, the Oak Valley community consists of a nucleus of around 80 people who live permanently at the site, plus a number of people who divide their time between Oak Valley and Yalata.[1]

The major logistic problem in sustaining the present-day community at Oak Valley is the lack of potable water.[12] Annual rainfall is low, there is no permanent surface water, and ground water is generally saline. Desalinated bore water was trucked 350 kilometres from Yalata during initial homeland development. From 1985 to 1988, Murray River water was transported by rail from Port Augusta to Watson and then trucked to Oak Valley. Ground water drilling programs in 1986 and 1987 resulted in several low-saline, low-yield bores in an area west of Oak Valley. Ecotech Pty Ltd was funded by the South Australian Health Commission during 1988–89 to investigate further options for the provision of water to Oak Valley.[12] A series of shed-tank units have now been installed within an approximate 100-kilometre radius of the community. The roof of each shed has a catchment area of 720 square metres and feeds tanks with a storage capacity of 72 kilolitres. The sheds are able to collect rainwater (especially from local thunderstorms) and were filled to capacity during good rains in 1988–89. The shed-tanks now provide almost all of the community's water needs and are more cost-effective than long-distance carted water.[12]

The establishment of further outstations on the Maralinga Lands at sites such as Muckera, the Wyola Lakes and in the vicinity of the Unnamed Conservation Park is under discussion.[11]

Tjuntjuntjara

As described in Chapter 8, the Aboriginal people of Cundeelee took over administrative control of the former mission in 1977 and formed their own incorporated body, Upurl Upurlila Ngurratja. The Western Australian Government acquired Coonana Station as a new community base and pastoral training centre for the Cundeelee residents in 1983.[13] The purchase did not achieve its intended aims because even greater problems were experienced with water supply, and only a small proportion of the station was capable of supporting pastoralism.[13]

In 1984, community elders expressed their disapproval by leaving Coonana and setting up their own camp north-east of the station on the Nullarbor Plain. Hasty government intervention resulted in the acquisition of 16,000 hectares of land at Yakatunya further to the north-east, but still on the northern margin of the Nullarbor Plain.[13] Yakatunya provided a base from which a return to traditional lands could begin.

A new homeland and resource centre was established at Tjuntjuntjara, 100 kilometres north of Yakatunya on the edge of the Great Victoria Desert, in 1988.[13] An incorporated body known as Paupiyala Tjarutja was formed a year later to represent the people of Tjuntjuntjara.[13]

With the help of consultant Scott Cane, Paupiyala Tjarutja have documented nearly 30 important traditional sites in the Great Victoria Desert north of Tjuntjuntjara. They include places of mythological, sacred and economic significance. Two sites have also been identified as possible locations for future outstations, namely Ilkulka, just south of the Anne Beadell Highway, and Tuwan, 75 kilometres north of Ilkulka. It is possible that Ilkulka will become the main service and resource centre for Paupiyala Tjarutja, with Tjuntjuntjara remaining a small population centre.[13]

Aboriginal access roads have been built through the heart of the Great Victoria Desert with Maralinga compensation monies.[13] Tjuntjuntjara is now linked with Ilkulka, Tuwan and Wingellina. Another track connects Tjuntjuntjara with Oak Valley and crosses the Unnamed Conservation Park. The purpose of these roads is to facilitate a resurgence in traditional cultural activity between closely related neighbouring communities.

Presently, Paupiyala Tjarutja has no legal tenure over its land. Tjuntjuntjara lies virtually in the middle of a rectangular extension to the western margin of the Great Victoria Desert Nature Reserve, while other proposed development locations are on vacant Crown Land to the north.[13] A proposal is currently before the Western Australian Government to consider the excision of the small rectangular block of land around Tjuntjuntjara from the Great Victoria Desert Nature Reserve.

Other Homelands

The Pitjantjatjara Lands now comprise a combination of major settlement bases (such as Ernabella and Amata), several relatively large communities which initially evolved from small homelands or enterprises (for example Pipalyatjara, Mimili and Fregon), and numerous 'satellite' homeland communities which feed off the larger population bases. Many Aboriginal people choose to remain at the major settlements because of their long residence and association with these settlements, and the ready availability of services.[3] Those persons in the satellite communities seek to combine the dual advantages of settlement and outstation life.[3]

Most of the small homeland communities in the Pitjantjatjara Lands are situated in or close to the range country but several, including Watarru (Mt Lindsay), lie on the northern fringe of the Great Victoria Desert.

A similar picture is presented in the Ngaanyatjarra Lands. After incorporation of an Aboriginal Council at Warburton in 1973, numerous small homeland communities were soon established along the range country to the east and north-east of Warburton. These included Irrunytju (Wingellina), Papulankutja (Blackstone), Mantamaru (Jameson), and Warakurna (Giles). Each had grown to a substantial community by the late 1980s. A network of satellite communities is now emerging around these bases. Yapupaara is at present closest to the Great Victoria Desert.

Mabo

On 3 June 1992, the High Court of Australia handed down a landmark decision on what has become known as the Mabo case. The full ramifications of the decision are yet to be determined. The High Court rejected the concept of *terra nullius*, that is the notion that Australia was an empty, unoccupied land in 1788 when European colonisation began.[14,15] It recognised for the first time the existence of indigenous title to land and, in the process, introduced a new term to Australian common law, namely 'native title'.[14]

The High Court ruled that Aboriginal people could claim native title and ownership over unalienated Crown Land, provided they had a demonstrable traditional connection

Anne Scott

LEFT: Irrunytju Community members continue to practise traditional hunting and gathering techniques in and around the Great Victoria Desert. Kuntjil Cooper gathers spinifex grass from the desert.

Anne Scott

ABOVE: Kuntjil Cooper winnows the spinifex grass.

Anne Scott

LEFT: Maisie Parker melts the spinifex resin.

Anne Scott

LEFT: The spinifex resin acts as an adhesive to repair the ends of this hardwood bowl (wira), which is used to hold water and collect berries and seeds.

BELOW LEFT: Polly Petermann is attempting to dig a perentie Varanus giganteus *from an underground rabbit warren.*

BELOW: Jean Burke proudly displays the captured perentie. The perentie inhabits rocky breakaways and sand dunes in the extreme western and north-western parts of the Great Victoria Desert.

Ken Scott

Ken Scott

with the land and, more importantly, they continued to acknowledge traditional law and observed traditional custom.[15] However, native title was extinguished when land had been alienated by the Crown (for freehold and most mining and pastoral leases).[14]

An 18-month period of volatile political and public debate followed the Mabo decision, before Federal Parliament finally passed its historic Native Title Bill in the early hours of 22 December 1993.[14] The *Native Title Act 1993* is neither land rights legislation nor a title granted by Australian courts or governments. It is a form of property right, governed by the laws and customs of indigenous people.[15]

The response of State Governments to the Commonwealth legislation has been varied. Western Australia will undoubtedly be the State most affected by the Mabo decision, and a large area of Crown Land in the Great Victoria Desert may qualify for native title. The Western Australian Premier, Richard Court, refused to support the federal *Native Title Act* and passed alternative legislation. Under his scheme, native title does not exist, but is replaced with a statutory right to traditional usage of land.[14] This enables Aboriginal people access to traditional lands, but does not allow them control over activities on the land.[14] The Western Australian Government intends to challenge the *Native Title Act* in federal court. In South Australia, the State Government under Premier Brown has tentatively supported the federal legislation, but has reserved the right to challenge it at a later time. Much of the South Australian side of the Great Victoria Desert is already under the tenure of Anangu Pitjantjatjara and Maralinga Tjarutja.

The relationship between native title and land reserved as National Park is somewhat nebulous. The High Court's judgement indicates that the creation of a National Park may in some instances not extinguish native title.[15] For example, if an area was vacant Crown Land before it became a park, native title may survive the creation of the park.[15,16] This would not negate the existence of the park, but would leave the relevant Aboriginal people with ownership rights that demand respect.[16]

The impact of Mabo on land tenure in the Great Victoria Desert cannot be fully determined until political and legal wrangling is resolved.

Chapter 12

Conservation of the Desert's Natural and Cultural Heritage

The Great Victoria Desert is one of the most aesthetic and unspoilt areas of Australia. Long parallel dunefields and vast sandplains characterise its landscape. Ancient palaeodrainage channels, saltlakes and claypans, undulating 'stony' plains, rocky breakaways and flat-topped hills represent a diverse array of other landforms found in the desert. The stately marble gum is the 'signature' tree of the Great Victoria Desert, while the Ooldea mallee, black oak, mulga and spinifex are other widespread and botanically interesting species. The desert's southern belt of mallee provides an important corridor for the movement of mallee-inhabiting birds and other fauna, while rarely-encountered birds such as the scarlet-chested parrot and Alexandra's parrot have been recorded in different habitats throughout the desert. The Great Victoria Desert is also one of the richest environments for reptiles in arid Australia and, indeed, the world. The desert's core is free from the impacts of pastoralism and devoid of introduced plants. As such, it provides a benchmark against which environmental degradation in other areas can be measured. There are powerful spiritual links between the desert and its Aboriginal custodians, who today continue to care for and manage the landscape as they have done for thousands of years prior to European contact. A network of traditional routes or 'highways', which are intimately associated with Dreaming tracks and key sources of water, cross the desert's landscape. Clearly, the Great Victoria Desert is a region that possesses exceptional natural and cultural values.

The Desert's Nature Reserves, Conservation Parks and Regional Reserves

As described in Chapter 11, much of the Great Victoria Desert is now under the tenure of Anangu Pitjantjatjara, Maralinga Tjarutja and the Ngaanyatjarra Council. In addition, a large proportion of the desert is afforded statutory protection through reservation as Nature Reserves (in Western Australia) and Conservation Parks and Regional Reserves (in South Australia). There are six A Class Nature Reserves on the Western Australian side of the desert, namely the Great Victoria Desert, Queen Victoria Spring, Neale Junction, Plumridge Lakes, Yeo Lake and De La Poer Range Nature Reserves (see Table 9 and the map on page 190).[1,2]

The Great Victoria Desert Nature Reserve encompasses nearly 2.5 million hectares of land adjacent to the Western Australian–South Australian border. The reserve provides

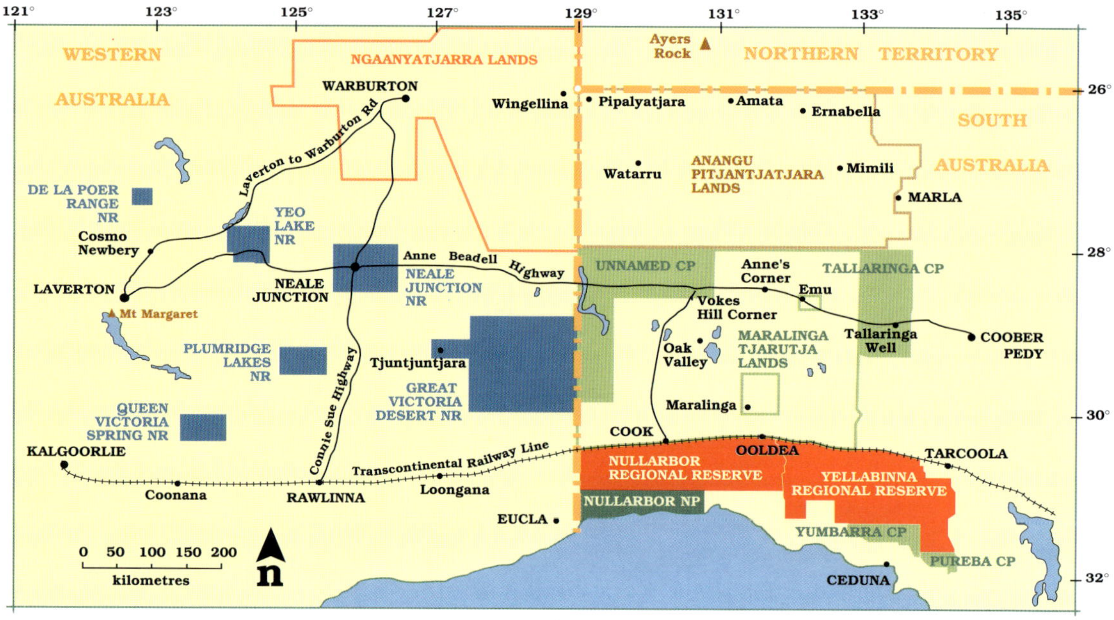

Conservation Parks, Regional Reserves and Native Reserves across the Great Victoria Desert.

protection for several important landforms and vegetation patterns of both the Nullarbor Plain and the Great Victoria Desert. The Nullarbor Plain, with its associated dongas and chenopod shrubland, covers the entire southern part of the reserve. Transitional myall woodland and bluebush form the major vegetation association in the reserve's centre. Dunefields more typical of the desert dominate the far northern section of the reserve. They support mallee eucalypt and spinifex communities and provide suitable habitat for a variety of fauna, including the marsupial mole.[1] The extensive Forrest Lakes system, as well as the Shell Lakes and Lake Ell, are also located within the reserve.

Queen Victoria Spring Nature Reserve lies in the south-western corner of the Great Victoria Desert and covers an area just more than 270,000 hectares. Queen Victoria Spring itself was of critical importance to its former Aboriginal occupants. The wider reserve protects representative areas of dunefield and sandplain, as well as granite outcrops, breakaways, large areas of undulating colluvial and alluvial plains, and a major palaeoriver (the Ponton Creek) with associated saltlakes and lunettes.[3]

The diversity of geological substrates and landforms, together with their derived soils, has resulted in a varied and interesting flora.[3] As mentioned in Chapter 2, the reserve's flora reflects an intermixing of elements from the Great Victoria Desert to the north-east and more mesic areas to the south and west. A number of regionally endemic species are also present.[3] The reserve contains one example of declared rare flora, *Conospermum toddii*.[3] Introduced plants and herbivores have had little impact. The sandhill dunnart and mulgara are important recent additions to the native mammal fauna of the reserve (see Chapter 4).

Plumridge Lakes Nature Reserve lies north-east of Queen Victoria Spring and is slightly larger in area than the former reserve. It is situated on the north-western

Andy Chapman

Mark Benbow, Mines and Energy SA

ABOVE: The Forrest Lakes are a major landscape feature of the Great Victoria Desert Nature Reserve.

LEFT: The Great Victoria Desert's dunefields often extend as far as the eye can see.

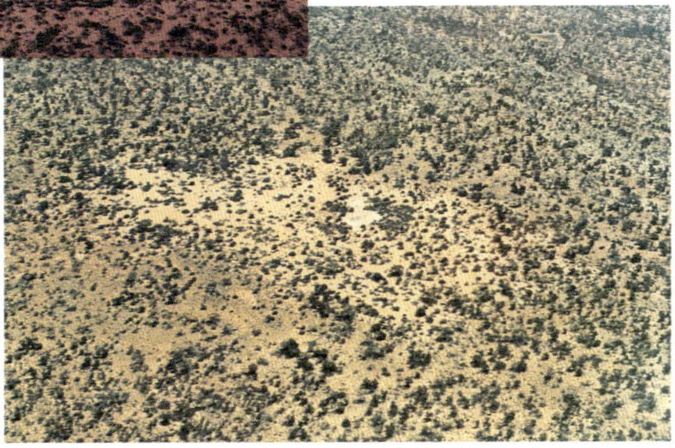

John Streich

RIGHT: Queen Victoria Spring, located in the middle of this aerial photograph, is an important cultural and historic landmark in the south-western corner of the desert.

TABLE 9: Parks and reserves in the Great Victoria Desert

State	Name	Classification	Year of proclamation	Area (hectares)
Western Australia	Great Victoria Desert	Nature Reserve	1970	2,495,777
	Queen Victoria Spring	Nature Reserve	1970	272,600
	Plumridge Lakes	Nature Reserve	1977	308,990
	Neale Junction	Nature Reserve	1977	723,073
	Yeo Lake	Nature Reserve	1979	321,946
	De La Poer Range	Nature Reserve	1991	75,000
South Australia	Unnamed	Conservation Park	1970	2,132,600
	Tallaringa	Conservation Park	1991	1,175,500
	Yellabinna	Regional Reserve	1990	2,522,700
	Yumbarra	Conservation Park	1968	327,589
	Pureba	Conservation Park	1990	144,400

boundary of the Nullarbor Plain and contains vegetation communities typical of both the Nullarbor fringes and the Great Victoria Desert.[1] Myall and bluebush dominate the south-eastern section of the reserve and provide suitable habitat for the Naretha blue bonnet. Black oak and mulga over bluebush occupy large sections of land west of Plumridge Lakes. Dunefields supporting marble gum and scattered Ooldea mallee are found in the north, north-western and western parts of the reserve. The scarlet-chested parrot has been recorded in the dunefield habitat.[4] The Plumridge Lakes and their associated saline flats occupy just over 11 per cent of the reserve area. Wombat tracks have recently been recorded around the shoreline of the lakes.[4]

Neale Junction Nature Reserve protects over 700,000 hectares of relatively pristine dunefields and sandplains which sustain extensive marble gum and hummock grassland communities. Areas of Ooldea mallee are also prevalent. Mulga woodland may be found in association with breakaway country in the northern part of the reserve. The scarlet-chested parrot has also been recorded in the Neale Junction area, while the reserve's overall sandy terrain supports a rich and diverse reptile fauna.

Yeo Lake Nature Reserve occupies just over 320,000 hectares of land in the central-western part of the Great Victoria Desert.[2] Approximately half of the reserve comprises sandplains, which support marble gums, Ooldea mallee and hummock grassland. Around 30 per cent of the reserve area contains relatively ungrazed mulga woodlands. The saline basin of Lake Yeo and its allied drainage channels occupy nearly 15 per cent of the reserve. A number of isolated breakaways and quartzite hills provide minor relief.[1] More than 75 species of birds are known from the reserve.

The recently dedicated De La Poer Range Nature Reserve lies in the far north-western corner of the Great Victoria Desert. It comprises three main landforms. The 'range' itself is, in reality, a large, eroded granite breakaway.[5] Away from the 'range', outwash areas are vegetated by mulga woodlands. Other parts of the reserve are dominated by sand dune/spinifex country, which are more typical of the desert. The granite breakaways have not been subject to grazing by sheep or goats and therefore provide a benchmark against which the impact of these animals can be measured.[5]

The six Western Australian Nature Reserves, although widely spaced, together aim to protect the complex compositional gradients found within the floral and faunal assemblages of the Great Victoria Desert. All the Nature Reserves are managed by the Department of Conservation and Land Management (CALM) and are vested in the National Parks and Nature Conservation Authority (NPNCA). CALM is about to release a Management Plan for the Goldfields Region. It is likely that this plan will contain new conservation and management initiatives for the Great Victoria Desert, in particular the Great Victoria Desert Nature Reserve. A future Nature Reserve is also proposed for the Lake Throssell area.

Four conservation parks and a regional reserve provide protection for the South Australian section of the Great Victoria Desert.

The Unnamed Conservation Park is a large, inverted 'L'-shaped park which encompasses 2.1 million hectares of the Great Victoria Desert and the northern limits of the Nullarbor Plain.[6,7] It is the largest Conservation Park in South Australia. The park adjoins the South Australian–Western Australian border and parts of the Great Victoria Desert Nature Reserve in the west. Its northern and eastern boundaries merge with the Aboriginal-owned Pitjantjatjara Lands and Maralinga Lands. The southern boundary of the park lies approximately 60 kilometres north of the Transcontinental Railway Line.

Most of the Unnamed Conservation Park comprises longitudinal dunefields and sandplains typical of the Great Victoria Desert. Many interdune corridors feature extensive surface exposure of calcrete and silcrete.[6,7] Open woodlands of marble gum, mallee, mulga and black oak with a varied shrub and grassland understorey dominate much of the park, while isolated pockets of native pine and kurrajong are also common. The ancient drainage channel of the Serpentine Lakes is a significant landscape feature along the park's western boundary. The highly nomadic Alexandra's parrot and rare scarlet-chested parrot have been recorded in the park. A number of important traditional water sources are located within the park's boundaries, while numerous Dreaming tracks criss-cross its terrain.

The unique natural and cultural values of the Unnamed Conservation Park have been recognised at a State and international level.[8] The fascinating history of the Unnamed Conservation Park and its pivotal role as a model for future conservation and land management practices are described more fully in a later section of this chapter.

Tallaringa Conservation Park is located approximately 100 kilometres due west of Coober Pedy and comprises just less than 1.2 million hectares of land on the eastern edge of the Great Victoria Desert. Prior to its dedication as a park in 1991, the Tallaringa region incorporated two separate parcels of land, each approximately 600,000 hectares in area.[9] The northern section was unallocated Crown Land. The southern block was held under annual grazing licence by Mabel Creek Station (Monta Rose Pty Ltd) from 1 September 1959 until 31 August 1989.[9] During 1988–89, the South Australian Pastoral Board decided that a future land-use assessment of both the northern and southern Tallaringa blocks should be conducted. Four separate reports were prepared by the Department of Lands and the Department of Environment and Planning, covering conservation, Aboriginal, pastoralism and mining interests.[9] The reports concluded that the Tallaringa Lands had high conservation and cultural values, but were incapable of sustaining a viable long-term pastoral enterprise without incurring substantial environmental impact. The area's potential for oil, gas and base

Mark Shephard

Andy Chapman

ABOVE: Neale Junction Nature Reserve lies in the heart of the Western Australian side of the Great Victoria Desert.

LEFT: Isolated breakaways are found in the Yeo Lake Nature Reserve on the desert's western edge.

Kathy Dynes (ANZ

RIGHT: Mulga woodlands dominate large sections of the Tallaringa Conservation Park.

metals was considered moderate to high. The entire Tallaringa region was subsequently declared a Conservation Park, following expiration of the grazing license over the southern block.

Tallaringa Conservation Park now protects nine major vegetation types not previously conserved within the Parks system in South Australia.[9] It also contains areas of regenerating mulga woodland that have not been subject to grazing pressure.[9] These woodlands provide a valuable benchmark for evaluating grazing impacts on adjacent pastoral leases (such as Mabel Creek, Commonwealth Hill, Mount Willoughby, Granite Downs and Wintinna West). The park also supports a relatively rich fauna for such an arid area. The Bourke's parrot is relatively common in the park, while the highly nomadic Alexandra's parrot was recorded during a recent survey.[9,10] The park also has immense cultural significance. Tallaringa Well is a site of major archaeological and mythological importance.[9]

The Yellabinna Regional Reserve comprises 2.5 million hectares of land south of the Transcontinental Railway Line. This reserve encompasses the south-eastern Great Victoria Desert and about half of the Yellabinna Environmental Association.[11] 'Regional Reserve' is a relatively new land classification in South Australia. It provides for the conservation of wildlife and the natural and historic features of the land while, at the same time, permitting pastoralism and the exploitation of natural resources.[12] In essence, it is a multi-use land management package.

Extensive tracts of mallee are found in the Yellabinna Regional Reserve. They form an important biological corridor connecting fauna from south-western Western Australia and the Eyre Peninsula in South Australia.

Mark Shephard

The Yellabinna Regional Reserve protects large expanses of relatively undisturbed wilderness. The area is a remarkable biological resource and is regarded as a stronghold for species such as the scarlet-chested parrot and sandhill dunnart.[11] Other interesting birds found in the region include the malleefowl, pink cockatoo, yellow-plumed honeyeater and yellow-rumped pardalote. The area also forms part of a key biological corridor connecting mallee woodlands in western and eastern Australia. Mt Finke is a locally interesting landscape feature which supports a high biological diversity of species.

As Copley states:

> *Yellabinna is... a key region for the conservation of biodiversity in its strictest sense; not just for the conservation of as wide a range of species as possible, but also for the conservation of as wide a range of genetic variation within the population of those species.*[11]

Two other Conservation Parks, Yumbarra and Pureba, adjoin the Yellabinna Regional Reserve at the extreme south-eastern tip of the desert. Both protect large areas of relatively pristine mallee. Yumbarra is one of the oldest conservation parks in South Australia. The endangered malleefowl has recently been recorded in this Park.[13]

The Unnamed Conservation Park — A Role Model for Future Land Management Practices

History of the Park

The concept of a large arid nature park in the far west of South Australia was first raised in 1967. At this time, the South Australian Pastoral Board examined the pastoral potential of vast tracks of land in the far west and north-west of the State (see Chapter 2). The Board's report concluded that the entire area south of the Anne Beadell Highway (latitude 28°30'S) possessed some potential for pastoral development. It also recommended that the area between 28°30'S and 28°07'S should be designated a Flora and Fauna Reserve to protect an area of marble gum, kurrajong and spinifex association. It suggested that southern parts of the North West Aboriginal Reserve could be added to the nature reserve to conserve a larger area. The report predicted confidently that the resultant Flora and Fauna Reserve would be 'unique, and second to none in Australia'.[14] Its attraction to tourists was deemed immense, while scientific research in the area would be of 'intense interest and value'.[14]

Acting in part on the Board's recommendation, over 2.1 million hectares of land in the Great Victoria Desert and northern Nullarbor Plain were subsequently dedicated as a National Park in 1970.[6] Its boundaries were drawn to include the western and southern extent of marble gum woodlands, the Serpentine Lakes palaeochannel, and myall and mulga communities on the northern edge of the Nullarbor Plain.[6] The park was not given a name. Two years later the park was reconstituted as the Unnamed Conservation Park, under the new *National Parks and Wildlife Act 1972*.[6] The 'unnamed' tag has stuck somewhat unimaginatively, despite several other names being proposed. Aboriginal people have numerous names for sites within the park, but no single name is representative of the area as a whole.[6] The Aboriginal word *Mamungarintja* refers to the Serpentine Lakes area and had been suggested as a possible new name for the park.[6] However, traditional owners have agreed that the park's official title should remain 'Unnamed'.

The international significance of the Unnamed Conservation Park was recognised in 1977, when it was proclaimed a Biosphere Reserve by the United Nations Educational, Scientific and Cultural Organisation's (UNESCO) Man and Biosphere Program.[6,7] It is now one of the largest arid-zone Biosphere Reserves in the world.

Biosphere Reserves aim to conserve the integrity of natural ecosystems and safeguard genetic diversity within those systems.[7] They also provide for scientific research and monitoring, public education, and a range of managed land uses including traditional practices.[6] Biosphere Reserves require the protection of a 'core' area of undisturbed terrain, together with surrounding 'buffer zones' which allow different types of land use.[7] In essence, Biosphere Reserves should be a model for the harmonious relationship between man and nature.[6]

The Unnamed Conservation Park readily fulfils many of the stated criteria for designation as a Biosphere Reserve. The park contains a cross-section of ecosystems representative of two significant arid landscapes, namely the Great Victoria Desert and the Nullarbor Plain. As such, it is a classic model of a natural biome.[6,8] It contains numerous examples of unique communities or areas with unusual natural features, such as marble gum woodlands, longitudinal dunefields and palaeodrainage channels of the Serpentine Lakes.[6,8] The park's remoteness and relative non-disturbance from introduced species ensure that it can be used effectively as a benchmark against which environmental impacts in more settled regions can be measured.[6] Its large area (2.1 million hectares) further guarantees that it can function as an effective 'conservation unit'.[8] It is also an example of a harmonious landscape resulting from traditional patterns of land use.[6,8] The entire park has been defined as a 'core zone', with the Anne Beadell Highway and Vokes Hill Corner to Cook Road classified as 'buffer or access' zones (see later).

In the late 1980s, the South Australian Government initiated moves to proclaim an integrated land and marine park over large sections of the Great Victoria Desert, the Nullarbor Plain and the near-shore waters of the Great Australian Bight. The latter are critical calving grounds for the southern right whale.[15] A three-stage proposal was put forward which involved cooperative arrangements with the Western Australian Government and Commonwealth Government and Aboriginal people with traditional affiliations to the area. Stage 1 of the plan focussed on South Australia. It recommended: (i) the acquisition of the Koonalda pastoral lease and part of the Yalata Reserve to form an extension of the Nullarbor National Park; (ii) the creation of a Regional Reserve over unallocated Crown Land north and east of the Nullarbor Park; and (iii) the transfer of title over the Unnamed Conservation Park to Maralinga Tjarutja who would then lease the park back to the South Australian National Parks and Wildlife Service. A marine park would also be created from the Head of the Bight to the South Australian–Western Australian border. In stage 2, the South Australian Government would seek a memorandum of understanding with the Western Australian Government for a consistent management plan to be developed for the Unnamed Conservation Park and adjoining Great Victoria Desert Nature Reserve. As a final stage, the South Australian Government would prepare a proposal for the Commonwealth Government nominating the area for World Heritage listing.[15]

The Australian Conservation Foundation (ACF) lauded the South Australian State Government for its initiative, and in April 1989 prepared a submission which

Mark Shephard

LEFT: Female southern right whales and their calves may be seen regularly off the Head of the Great Australian Bight in winter. During the 1980s the South Australian Government and the Australian Conservation Foundation proposed to include the Great Victoria Desert, the Nullarbor Plain and the waters off the Head of the Bight in an integrated Land and Marine Park.

Chris Wenham

RIGHT: The Serpentine Lakes in the Unnamed Conservation Park.

recommended a slightly modified area for World Heritage nomination. The ACF suggested that the southern boundary of the Unnamed Conservation Park should be extended some 290,000 hectares south to meet the Transcontinental Railway Line. The Western Australian Government were to be encouraged to unify under conservation tenure adjoining parts of the Great Victoria Desert and Nullarbor Plain, and make them continuous with the South Australian reserves. A totally integrated land and marine park could thus be created, which the Foundation believed would strengthen immeasurably later arguments for World Heritage listing. The proposed land package would represent the largest terrestrial National Park in Australia.[15] The three governments were to enter into a joint management agreement for the land and marine park, with the Commonwealth Government providing funding and personal support and the State Governments bearing responsibility for day-to-day park management. The provision for stewardship of Aboriginal people over their traditional lands was seen as a major trump card in meeting requirements for World Heritage listing.[15] Unfortunately, the ACF proposal lapsed in the ensuing years as a result of difficulties in achieving universal intergovernmental support.

In 1992, the South Australian Government passed new legislation entitled the *Wilderness Protection Act*. The Act provides a mechanism by which wilderness areas may be identified, assessed, proclaimed and managed.[16,17] The Act also acknowledges that there is little, if any, pristine wilderness left in Australia. It therefore seeks to evaluate the degree to which wilderness qualities of remoteness to access and settlement, and aesthetic and biophysical naturalness have been modified by social and technological development.[17–19] Under the Act, the government-appointed Wilderness Advisory Committee is charged with the task of identifying those areas of the State which satisfy wilderness criteria to an extent sufficient to justify their protection under the Act.[17]

Two types of wilderness classification are embodied within the Act. Wilderness Protection Areas may be proclaimed over all or part of a reserve constituted under the *National Parks and Wildlife Act 1972*, Crown Land or any other land with the owner's consent.[17] Wilderness Protection Zones are a secondary category, established to provide a mechanism for a wilderness management framework over land where there is an existing mining tenement.

A Wilderness Code of Management establishes principles for the protection of wilderness values, ecosystems, flora and fauna, Aboriginal and non-Aboriginal heritage, as well as for management of visitors, scientific research and fire in wilderness protection areas and zones.[17]

Five areas on Kangaroo Island have been recommended initially for protection under the Act. The Unnamed Conservation Park has unquestioned wilderness quality and is currently under assessment by the Wilderness Advisory Committee.

The somewhat romantic concept of wilderness has been part of white Australian culture for most of this century. However, the concept of wilderness is fundamentally alien to Aboriginal Australians.[6] Aboriginal people do not think in terms of 'wilderness' but rather 'country', that is, particular landscapes within which its people have played an integral role for around 50,000 years.[20]

Joint Management of the Unnamed Conservation Park

A Management Plan for the Unnamed Conservation Park was formally adopted in 1993.[6] The plan recommends that Maralinga Tjarutja be granted inalienable title to the park, and that the park be jointly managed by the South Australian Department of Environment and Natural Resources (DENR) and its traditional owners.[6] This type of management arrangement has proven successful in several parks in the Northern Territory (the most well-known example being Uluru), but creates a precedent in South Australia. Amendments to the *National Parks and Wildlife Act 1972* facilitating title transfer of reserves to Aboriginal bodies were introduced into the South Australian Parliament in October 1992.[21] The amendments were subsequently withdrawn and have been held in abeyance, pending a comprehensive review of the National Parks system in South Australia.[21] This review was completed in mid-1994 and concluded that 'the aspirations of Aboriginal people to resume ownership and share management of reserves need not be in conflict with the ongoing nature conservation management of such reserves'.[21] Indeed, joint management would provide a positive form of reconciliation between black and white Australians and create a mechanism by which traditional Aboriginal knowledge could be integrated into existing reserve management.

To Aboriginal people, joint management (or *tjungaringu* in Pitjantjatjara) means the park has two 'bosses' who contribute equally to management decisions.[6] *Tjungaringu* carries the responsibility of full commitment, honesty and integrity by both partners.[6,22]

Joint management of the Unnamed Conservation Park will function by way of a formal Board of Management.[6] The six-member Board will include an Aboriginal Chairperson from Maralinga Tjarutja and three traditional owners nominated by Maralinga Tjarutja. A 'traditional owner' is an Aboriginal person who, in accordance with

Aboriginal tradition, has social, economic and spiritual affiliations and responsibilities for the park.[6]

Archie Barton, Maralinga Tjarutja's Administrator, has been a major driving force behind the cooperative spirit that now exists between DENR and the park's traditional owners. He was born at Barton railway siding in the south-eastern Great Victoria Desert. Barton was taken from his mother at an early age and removed to a children's home in Port Augusta. He saw his mother only once more before she died. Barton has been a pioneer and leader in bridging gaps between black and white Australians and was awarded an Order of Australia in 1989.

Conservation by Colours — Green and Black Working Together

Conservation of the environment has been an important and emotional issue among white Australians for most of this century. The 'green' conservation movement has become a major political force in recent years. However, the European record of land management is certainly not without tarnish. Future generations of Australians are now left with legacies of wholesale land clearance, soil erosion and salinity, pollution of air and waterways and the impact of feral species.[23] Aboriginal people have enjoyed a more harmonious relationship with the land extending over 50 millennia, although it must be acknowledged that traditional burning practices have undoubtedly reshaped the arid landscapes of Australia's deserts. Aboriginal interests in nature conservation have nonetheless been largely ignored by white Australians until the 1980s; furthermore, their rights to ownership of land (including conservation parks and nature reserves) were generally dismissed and most parks/reserves were created without the involvement or consent of Aboriginal people.[23]

On 7–9 June 1988 a landmark conference was held at Emu on the role of Aboriginal people in nature conservation and land management.[24] The conference was hosted by the Oak Valley Community on behalf of Maralinga Tjarutja. It was attended by more than 200 Aboriginal people representing more than 20 communities/councils from South Australia and Western Australia, as well as by members of State and federal conservation agencies, related government organisations and community-orientated conservation groups. A series of workshops were held on topics such as feral animal control, reintroduction of endangered species, vegetation loss, soil erosion and tourism.[24]

The conference provided the foundation for cross-fertilisation of European and Aboriginal land management principles. State and federal representatives were able to listen to discussions on land management issues in an Aboriginal context on Aboriginal land.[24] The conference reinforced Aboriginal people's strong desire to be directly involved in the management of land (including conservation parks and nature reserves), and to develop employment opportunities from such involvement.[24]

The implementation of subsequent joint management schemes for the Unnamed Conservation Park and other Parks around Australia, with their inherent recognition of rights for traditional ownership of land, has provided an excellent model for reconciliation of black and green conservation interests.

It is likely that a joint management scheme will be undertaken for Tallaringa Conservation Park in the near future and involve the recently formed Tjaliri Aboriginal Corporation. The Department of Conservation and Land Management (CALM) in

TABLE 10: Summary of data from the Visitors Book at Vokes Hill Corner in the Unnamed Conservation Park

Year	No. of vehicles	No. of visitors	Earliest entry for the year	Latest entry for the year
1987	58	95	11 March	29 October
1988	89	246	7 February	7 November
1989	97	226	14 February	29 November
1990	99	235	30 March	9 November
1991	167	367	15 April	31 October
1992	121	292	4 April	9 October
1993	118	328	23 March	19 November

Data on number of vehicles and visitors were compiled by Pia Richter, Ranger, DENR Far West District Office, Ceduna.

Western Australia has recognised that traditional ecological knowledge can play an important role in management of the Great Victoria Desert Nature Reserve. It is now actively seeking and encouraging joint liaison with Paupiyala Tjarutja in field work and research activities being undertaken in the reserve.[25]

As mentioned in Chapter 11, the relationship between the Federal Government's *Native Title Act* and land reserved as National Park is somewhat unclear. The High Court's judgement indicates that the creation of a National Park may in some instances not extinguish native title.[26] For example, if an area was vacant Crown Land before it became a park, then native title may survive the creation of the park.[26,27]

Rather than being confrontationist, the *Native Title Act* simply recognises a property right of Aboriginal people that has always existed. Taken in this spirit, the Mabo decision can be a springboard for reconciliation between black and white Australians. Conservation management has the opportunity to be at the leading edge of this process.[28] Green conservationists must come to terms with the special relationship Aboriginal people have forged with the land and the complementary skills they bring to land management. There is considerable common ground between black and green conservationists in what they desire to protect, albeit for slightly different reasons.[27] There is no doubt that Aboriginal culture and the future of the Australian environment are inextricably and irreversibly linked.[23,28] Land aspirations of both groups can be combined for the betterment of all Australians.

Land Management Issues in the Great Victoria Desert

Visitor Access

The Great Victoria Desert is becoming an increasingly popular destination for four-wheel drive adventurers and provides outstanding recreational opportunities. Because of the remoteness of the park, most visitors travel in small, self-sufficient convoys or as part of a trip organised by a four-wheel drive club or licensed tour operator. Figures from the visitor's book at Vokes Hill Corner in the Unnamed Conservation Park indicate that around 120 vehicles now pass through the area annually, most during the winter months (see Table 10).

The major approved access roads across the desert are the Anne Beadell Highway, the Connie Sue Highway, the Vokes Hill Corner to Cook road and the Laverton to Warburton road (see Appendix). The Anne Beadell Highway is the main west–east route through the desert. The Connie Sue Highway runs in a north-south direction from Warburton to Rawlinna. The Vokes Hill Corner to Cook road provides alternate access to the Unnamed Conservation Park. The Laverton to Warburton Road crosses the north-west corner of the Great Victoria Desert. A number of different permits MUST be obtained prior to traversing these main access roads (see Appendix).

There are also other roads across the desert which are for Aboriginal access only. These include the Great South road from Wingellina to Tjuntjuntjara, the road linking Oak Valley and Tjuntjuntjara, the road from Anne's Corner to Pipalyatjara, and several others. Non-Aboriginal visitation along these roads is not permitted, unless prior written approval is obtained from the relevant Aboriginal Council/Board of Management (see Appendix).

The recently adopted Management Plan for the Unnamed Conservation Park defines several categories of Access Zone and identifies specific management objectives for these zones.[6] The Anne Beadell Highway is encompassed within the Northern Access Zone, which is 4 kilometres wide except at the Serpentine Lakes where it widens to 10 kilometres. Camping is permitted within 100 metres of the road, except within a 50-kilometre section of the highway between longitudes 129°20' and 129°48'. This 'no camping zone' has been designated because it is near highly sensitive traditional sites. The Northern Access Zone will be managed to 'provide opportunities for visitors to appreciate the Park's natural and cultural values, while at the same time preserving the remote and undisturbed character of the Park' and to 'maintain Aboriginal tradition and facilitate the social and economic development goals of traditional owners'.[6]

A 20-kilometre wide Southern Access Zone has been declared in the south-western corner of the park, where several tracks lead from Hughes on the Transcontinental Railway Line into the area.

A Central Access Zone has been created around a road which crosses the park and links Oak Valley in South Australia to Tjuntjuntjara in Western Australia. This road has been specifically built to facilitate the re-establishment of cultural ties between the two closely related Aboriginal communities.[6]

All visitors to the Great Victoria Desert have a strong obligation to minimise their impact on the environment and to respect and care for its unique landscape. The Friends of the Great Victoria Desert Parks was formed in March 1994.[29] This group provides voluntary support and assistance to National Parks and Wildlife Service rangers responsible for the parks and reserves in the South Australian section of the desert. It is one of more than 75 similar Friends of Parks' groups in South Australia and gives an opportunity for members of the general public to make a positive contribution to the conservation of a desert environment.

Wildlife Management

Protection and preservation of the Great Victoria Desert's native plant and animal communities is a significant conservation objective. Ongoing scientific research and

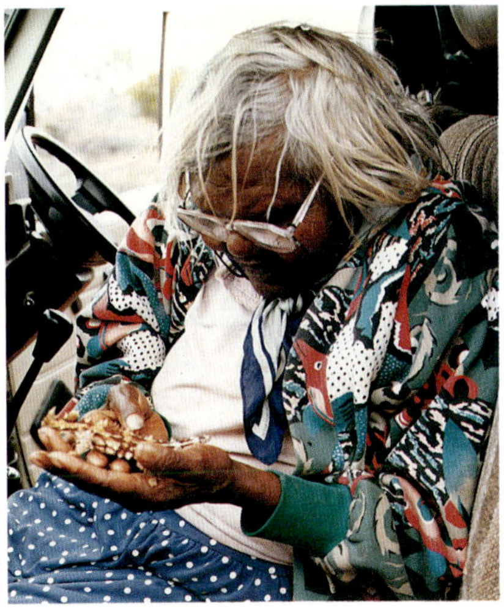

Kathy Dynes (ANZSES)

Mark Shephard

Kathy Dynes (ANZSES)

LEFT: Jessie Lennon, a member of the Tjaliri Aboriginal Corporation, examines a thorny devil as part of the biological survey work conducted by ANZSES in the Tallaringa Conservation Park during October 1993. Traditional Aboriginal knowledge can often enhance or provide an alternative perspective to modern scientific research.

TOP RIGHT: Members of the Friends of the Great Victoria Desert Parks met Maralinga Tjarutja elder Peter Pepper (standing, second from left) and park ranger Albert Zepf (standing, third from left) near Vokes Hill Corner in September 1994.

RIGHT: Museum specimens are an important identification aid for scientists and amateur naturalists working in the Great Victoria Desert.

monitoring, traditional ecological knowledge and data collection by 'amateur' naturalists can all contribute to a better understanding of the desert's wildlife populations. No known living species found in the Great Victoria Desert is in present danger of extinction. However, complete inventories of the desert's flora and fauna need to be maintained and monitored to ensure events such as the dramatic decline that has occurred in the desert's mammal population are not repeated. Complacent attitudes to wildlife management should not be tolerated.

The remoteness of the desert has, to a large extent, assured its biological integrity. Introduced plants are virtually absent from the core of the desert, although their presence is increasing around the desert's fringe. The potential for exotic plant invasion is ever present, however, and is greatly enhanced by increased vehicular movement through the desert.

Feral animals are firmly established in the desert. The European rabbit poses a major potential threat to the desert's ecology. Its burrowing activities expose delicate topsoil to erosion, its voracious appetite suppresses regeneration of vegetation communities, and its sheer numbers overwhelm native species competing for similar niches. The house mouse is widespread throughout the desert and may irrupt to plague proportions under favourable conditions. Cats and foxes are skilled hunters of native reptiles, birds and mammals. The camel is now entrenched as part of the desert's fauna. Its insatiable appetite has an impact on vegetation biomass, while its capacity to drink waterholes dry can cause devastating effects for local native fauna. Regular monitoring of all feral animal populations should be conducted across the desert.

In the Unnamed Conservation Park, traditional owners are allowed to hunt native fauna and gather native plants for food or cultural purposes under a special proclamation of the *National Parks and Wildlife Act 1972*. Some conservationists argue that hunting of native animals, particularly the endangered Australian bustard, is incompatible with the park's management objective of wildlife preservation. In practical terms, the scale of hunting in the park is small and its overall impact is considered negligible.[6,22] Of potentially greater concern is the method by which hunting is conducted. Traditional hunting on foot and with spears has largely been replaced by the use of European vehicles and firearms. Indiscriminate cross-country four-wheel drive travel to reach traditional hunting areas may cause considerable environmental damage. Aboriginal people, like European visitors, have a responsibility to minimise vehicular impact.

Protection and Monitoring of Cultural and Historic Sites

Another important conservation objective for the desert should be the protection, maintenance and regular inspection of sites of European or Aboriginal heritage.

The signposts erected by Len Beadell throughout the desert are of historic value, but some have been subjected to needless vandalism or physically removed by 'souvenir' hunters. There are literally hundreds of important Aboriginal sites throughout the desert. Many sites are currently under the care and custody of Aboriginal people. However, other sites lie on Crown Land and are not subject to any formal management or protection plan.

Fire Management

Aboriginal patch-burning techniques have been widely practised in the Gibson Desert and Central Australia as part of integrated fire management strategies designed to maintain habitat and species diversity and to restrict the spread of wildfires. Conservation and Land Management (CALM) authorities have adopted a similar proactive burning policy and maintain a controlled annual burning program on the Western Australian side of the Great Victoria Desert.

In South Australia, a study was conducted into the recent fire history of the Unnamed Conservation Park and surrounding areas of the Great Victoria Desert during 1992.[7] It revealed that natural wildfires occurred infrequently and were mainly confined to the northern section of the park where highly flammable spinifex was prevalent. The exception to this generalisation was a series of major fires in 1974–75 which burnt around 19 per cent of the park. These fires were fuelled by an abundant growth of herbs and grasses which followed exceptional rains during the previous summer.

The study concluded that a single wildfire at any point in time would be unlikely to totally burn the entire park or a particular vegetation community because of the park's vast size and wide distribution of major plant associations. As such, the development of a proactive burning strategy to maintain natural systems and traditional fire regimes was considered inappropriate for the Unnamed Conservation Park.[7] Further, the suppression of wildfire in the park was deemed unfeasible because of both its inaccessibility and the cost involved in bringing fire-fighting teams and equipment to the area.[7]

Pastoralism

Pastoralism is not a major land use in the Great Victoria Desert. On the South Australian side of the desert, the Far West, Far North West and Tallaringa regions have been assessed for their pastoral potential.[9,14,30] Each area was considered capable of supporting stock to varying extents, with land south of latitude 28°30'S having the potential to maintain 600,000 sheep. A combination of factors including high capital costs, predicted low and erratic returns, lack of water supplies, and the establishment of various Conservation Parks has precluded any further development of the pastoral industry in these areas. Pastoralism has continued to be practised at a number of stations on the desert's eastern fringe including Mabel Creek, Mt Willoughby and Commonwealth Hill. The Pitjantjatjara Council is also investigating the possibility of re-establishing cattle in sections of the northern Anangu Pitjantjatjara Lands.

In Western Australia, pastoral leases are held over land along the southern and western fringes of the Great Victoria Desert (for example at Seemore Downs, Yamarna and Lake Wells).

Mining

The exploration for minerals and petroleum is subject to stringent controls over much of the Great Victoria Desert.

In South Australia, mining is prohibited in the Unnamed Conservation Park, except by a resolution of both Houses of State Parliament.[6] A joint proclamation under Section 43 of the *National Parks and Wildlife Act* allows mining in Tallaringa Conservation Park, following approval by the Minister of Environment and Natural Resources.[31] Mining is legally permitted in the Yellabinna Regional Reserve under amendments to Section 34 of the Act.[31]

Exploration in the Anangu Pitjantjatjara Lands and Maralinga Tjarutja Lands is subject to provisions under existing Land Rights Acts.[32–34] A company wishing to obtain a petroleum or mining tenement within the South Australian Aboriginal Lands must first apply to the State Minister for Mines and Energy. Following Ministerial approval, the company may then begin negotiations with the Aboriginal body to determine terms and conditions under which it can enter the Lands. The Minister will then grant a tenement to the company, provided satisfactory agreement on entry conditions has been reached. A mechanism of appeal is available under the Act, where access negotiations have broken down. The Aboriginal bodies receive one-third of the total royalties payable from mineral or petroleum resources subsequently developed on their Lands.[32,33] Some minor differences exist between the two Land Rights Acts. For example, the *Maralinga Tjarutja*

Act enables the recovery of reasonable costs and expenses incurred in dealing with entry applications, but limits the payment of compensation for subsequent disturbance to the Lands, their people and their way of life to an amount no greater than that provided for under current Mining and Petroleum Acts.[32,33]

In Western Australia, an exploration licence may be granted in any of the Great Victoria Desert's six A Class Nature Reserves, subject to the concurrence of the State Minister for the Environment.[35] Before an application is considered for such reserves, the intending explorer must provide details of the proposed exploration program including environmental management. This allows the Department of Conservation and Land Management (CALM) and the National Parks and Nature Conservation Authority (NPNCA) to review the application before making a recommendation to their minister. The conditions applied to such exploration are very stringent and there is a stipulation for ongoing reporting of the progress of exploration and rehabilitation. Mining proposals for A Class Nature Reserves also require assessment by the Environmental Protection Authority (EPA) and the consent of both Houses of State Parliament.[35]

Exploration licences on vacant Western Australian Crown Land include conditions that require the agreement of the State Department of Minerals and Energy before there is any ground-disturbing activity using mechanised equipment.[35]

Mining activities in the Great Victoria Desert's Western Australian Aboriginal Reserves are subject to the provisions of two statutes: the *Mining Act 1978* and the *Aboriginal Affairs Planning Authority Act 1972*.[35]

Under the *Mining Act 1978*, the Minister for Mines may grant a mining title and written consent to mine, following the recommendation of the Minister for Aboriginal Affairs. An entry permit is also required from the Minister for Aboriginal Affairs to gain access to a Reserve. Before making the recommendation or issuing the entry permit, the Minister for Aboriginal Affairs must consult with the Aboriginal Lands Trust as the vesting body for the land. The Trust, in turn, consults with the relevant Aboriginal body or community (for example the Ngaanyatjarra Land Council).[35] The conditions of the entry permit are formulated on the basis of an agreement reached between the tenement holder, the relevant local body and the Lands Trust. A model agreement, which covers issues such as Aboriginal heritage protection and compensation, has been developed to facilitate and streamline the negotiation process as much as possible.

Equivalents of rents and royalties are collected by the State Department of Minerals and Energy following mining activities on Western Australian Aboriginal Reserves. They are paid by the State Treasury to the Aboriginal Lands Trust, which may then allocate some of the funds to the relevant Land Council or local community.[35]

Very little exploration activity has taken place on Western Australian Aboriginal Reserves in the Great Victoria Desert since the early 1970s. However, in response to the recent streamlining of relevant administrative procedures, increased mining interest within the Ngaanyatjarra Lands and Cosmo Newbery Aboriginal Reserve is likely.

The petroleum potential of the Officer Basin, which underlies much of the Great Victoria Desert, was first examined as far back as the early 1950s.[36] In South Australia, oil exploration licences from 1958 to 1974 were held mainly by Exoil and Outback Oil.[32] Seismic, aeromagnetic and gravity surveys were conducted during this period. Numerous wells were also drilled, including several near Emu and one north of the Unnamed Conservation Park (Birksgate Well No. 1).[33]

A prescribed spring fire in the Queen Victoria Spring Nature Reserve is designed to break up large areas of uniformly aged spinifex to limit the spread of future fires and to enhance wildlife habitats.

Comalco carried out extensive petroleum and mineral exploration programs on the South Australian side of the basin from 1980 to 1989.[32] Petroleum exploration licence number 23 (PEL 23) was granted to Comalco in January 1983 and included a large block of land north of Emu in the north-east corner of the Maralinga Lands.[32,33] PEL 29 was secured over an area in the south-eastern Anangu Pitjantjatjara Lands in November 1985 by a consortium of companies including Amoco Australia Petroleum Company and Anangu Pitjantjatjara Oil Pty Ltd (a company owned by the Pitjantjatjara people). Both licences were relinquished in the late 1980s, mainly as a result of a downturn in the oil industry.[33]

During 1993 a major seismic survey was conducted in the central Officer Basin by the Australian Geological Survey Organisation (AGSO), formerly the Bureau of Mineral Resources.[37] The survey formed part of the Geoscience Mapping Accord. The South Australian Department of Mines and Energy provided logistical assistance (Aboriginal liaison, surveying and water bore drilling) using funds from the South Australian Exploration Initiative.[32] The object of the survey was to develop an improved understanding of the structure and stratigraphy of this largely unexplored portion of the basin and to encourage further petroleum exploration. More than 620 kilometres of seismic lines were constructed including a single line that commenced near Mt Lindsay, passed through Birksgate Well No. 1, and stopped at the northern boundary of the Unnamed Conservation Park. Four seismic lines were made south and east of the park. These included two which directly abutted and ran parallel to the park's boundaries, and another which was made along an existing section of the Vokes Hill Corner to Cook road.[37]

The South Australian Department of Mines and Energy called for licence applications over four areas within the eastern Officer Basin in 1994.[34] These new areas encompassed sections of land formerly covered by licence numbers 23 and 29, as well as an area north of Barton in the south-eastern Maralinga Lands. The future potential for discoveries of oil in this part of the Officer Basin is regarded as high.[36]

In Western Australia, the Bureau of Mineral Resources, Geology and Geophysics (BMR) undertook reconnaissance geological mapping in the western Officer Basin during the 1950s and 1960s.[38] They also conducted limited aeromagnetic, gravity and seismic surveys. A detailed petroleum exploration program was subsequently mounted in the early 1960s by a consortium headed by the Hunt Oil Company.[38] Part of their exploration focussed on the northern Great Victoria Desert. Further regional geological mapping of the basin was conducted by BMR and the Geological Survey of Western Australia (GSWA) during the 1970s. More recently there has been renewed interest in petroleum exploration in the western Officer Basin. The Shell Company of Australia was granted three petroleum exploration permits in 1980; the area of land covered by these permits included part of the Ngaanyatjarra Lands in the north-western Great Victoria Desert.[39]

There are numerous other potential natural resources in the Great Victoria Desert apart from petroleum. In South Australia, these include base metals, uranium and gold in ancient rocks of the Gawler Craton and older sediments of the eastern Officer Basin; coal in the Arckaringa Basin; and diamonds and opals in the desert's younger Mesozoic and Cainozoic sediments. There is also potential for finds of heavy minerals in the palaeobeach sands of the Barton and Ooldea Ranges, and uranium and evaporites in the desert's palaeochannels and saltlakes.[36]

Stockdale Prospecting conducted extensive diamond exploration in the Maralinga Lands north of Barton between 1981 and 1987.[36] BHP commenced a mineral exploration program in the Ooldea Range during 1990, following an agreement on site protection with the area's traditional owners.[33] The South Australian Department of Mines and Energy conducted a bedrock drilling program in the Barton block between June and December 1993 to improve understanding of the area's basement geology and to increase its mineral prospectivity.[40]

In Western Australia, the desert's most significant minerals include gold and copper deposits in greenstones in the eastern Yilgarn Craton; the Mulga Rock and Thatcher Soak uranium deposits; and the Yamarna chromite deposits.[41,42] Substantial gold and nickel production has occurred in the vicinity of Laverton, just outside the western edge of the Great Victoria Desert.[41] The Mount Weld phosphate and rare earth deposit is also located near the western boundary of the desert.[43] CRA Exploration Pty Ltd, Ashton Mining and Western Mining have all recently carried out exploratory work for diamonds in the north-western and central-western Great Victoria Desert.

Guidelines on environmental management and field work practices within conservation parks and nature reserves have now been developed in both South Australia and Western Australia.[44,45] The South Australian guideline document results from careful negotiation and agreement between the National Parks and Wildlife Service, the Aboriginal Heritage Branch of the Department of Environment and Planning, the South Australian Department of Mines and Energy and the Chamber of Mines and Energy. It describes procedures to minimise soil, vegetation and fauna disturbances, and groundwater contamination and pollution. It also details formal consultative processes to ensure all documented heritage sites within exploration areas are known. Recommendations on rehabilitation and stabilisation of disturbed areas, and monitoring and inspection of ongoing field work practices are also provided.[44]

Appendix

TRAVELLING THE GREAT VICTORIA DESERT

The Great Victoria Desert is a place for the dedicated outback traveller. Adequate time is needed to fully appreciate, enjoy and experience the contrasting moods and character of this delightful wilderness, to unwind from the pressure of everyday living, and to adjust to nature's pace of life. The visitor to the Great Victoria Desert will also need to be well prepared and totally self-sufficient, as it is more than likely that no other human presence will be encountered during a journey across the desert.

The following information is presented as a **guide** for the intending desert traveller.

Plan the Trip Carefully

The importance of careful and meticulous planning of a trip cannot be overemphasised. It is important to read as much literature on the desert as possible prior to departure, as this will provide the visitor with a better appreciation of the desert's features and points of interest (see Bibliography). Talk to friends and acquaintances who have previously driven your intended route to gain basic information on track conditions and travelling times.

Owing to the remoteness of the desert, plan to travel in the company of at least one other vehicle. Never attempt to visit the desert alone. A convoy of three to four vehicles is considered ideal. Travelling in convoy affords a number of advantages, such as minimising the risk of becoming stranded in the event of breakdown or accident, allowing for load sharing, and creating a greater peace of mind from a feeling of 'safety in numbers'. It is imperative that at least one member of the travelling group possesses sound mechanical and navigation skills and knowledge of first aid.

A number of permits will be required for visiting particular parks and Aboriginal land in the Great Victoria Desert (see later). The task of applying for permits should be commenced at least three months prior to departure. Aboriginal Councils generally hold only one meeting a month to consider permit applications. If an application is received just after such a meeting, then there can be a minimum of two months' delay in finalising permits. Permit applications should include the names and addresses of all members of the party, vehicle makes and registration numbers, details of your intended route, and dates of entry and departure from each area.

A convoy of three to four vehicles is the safest, most practical and sensible way to travel the Great Victoria Desert.

The best time to visit the Great Victoria Desert is in the cooler months of the year, between April and September. Days are generally fine to mild during this period, although nights may be very cold (around 0°C or below) in mid-winter. Extremely hot day temperatures of up to 50°C may be encountered outside the suggested visitation period.

A master list of supplies and equipment should be prepared initially, and then refined and updated regularly. Do not wait until the last day before seeing if everything will fit into the vehicle. It is advisable to have several 'practice packs' a week or two before departure to ascertain just how much space is (or, more importantly, isn't) available. As a general rule, the lighter the vehicle, the lower the fuel consumption and the less wear and tear on the vehicle. Overloading is a major cause of vehicle failure. Equipment should be arranged according to bulk, weight and accessibility. Heavy items should be kept low and forward in the vehicle. An overloaded roof rack should be avoided as it increases the risk of vehicle roll over. A mesh cage barrier is a practical means to prevent a load falling forward. Many items can be conveniently stored in stackable, lidded crates, while any loose items should be securely fastened. Soft materials, such as pillows and rugs, can also be used effectively to hold loose material in place.

Food, fuel and water requirements should be carefully considered. There are no services or facilities available along the entire length of the Anne Beadell or Connie Sue Highways, for example. The need to be totally self-sufficient is therefore paramount.

Mark Shephard

Reaching the crest of a dunefield and looking at the view ahead is an exciting experience for the desert traveller.

A car fridge is an essential item for the desert traveller. A variety of foods can be taken on your trip, including barbecue meats, camp-oven roasts, vegetables, fruit, health bars, pancake mix, and assorted tinned or packet desserts, to mention but a few. Do not pack food or drinks in glass containers because glass may easily shatter while traversing uneven track surfaces. One of the great delights of desert travelling is consuming a hearty evening meal, cooked around the campfire under a sky filled with flickering stars.

A water allowance of 6 litres per person per day is an approximate guide to the quantity of this fluid that will generally be used for drinking, cooking and washing purposes.

Fuel requirements need to be carefully calculated, and will vary depending on factors such as type and age of vehicle, weight carried, tyre pressure and track conditions. A six-cylinder, long wheel-base, diesel vehicle should have an average fuel consumption of around 10–15 litres per 100 kilometres during a desert crossing. A six-cylinder, long wheel-base vehicle using unleaded petrol should average between 15 to 22 litres per 100 kilometres. It is critical that drivers calculate their individual fuel requirements, based on the total distance to be travelled and the known average fuel consumption for their own vehicles. In addition, it is strongly recommended that an extra 40–50 litres of fuel be carried in reserve, in case of vehicle breakdown, misjudgement of track surfaces or unforeseen detours caused by unseasonal weather.

As a working example, if a diesel vehicle with a fuel consumption of 12 litres per 100 kilometres is to be taken on a trip covering a distance of 1500 kilometres, then a total of 180 litres plus 40 litres reserve, that is 220 litres, should be carried.

It is important to take appropriate navigation/communication equipment. At least one vehicle in the travelling party must carry a radio capable of operating on HF Royal Flying Doctor Service or HF Telstar frequencies. All vehicles should have similar short-range radios (for example UHF 470 mHz or Citizen Band 27 mHz) for inter-vehicle communication. Global positioning systems (GPS) are becoming increasing popular among desert travellers, but remain very expensive working tools. The following 1: 1,000,000 topographical maps cover the Great Victoria Desert and surrounding areas: Wiluna, Kalgoorlie, Petermann Ranges, Nullarbor Plain, Oodnadatta and Tarcoola. Also available for more detailed examination of particular areas are 1: 250,000 maps .

Vehicles should be thoroughly serviced at least three weeks prior to departure to ensure that they are in first-class mechanical condition. A comprehensive range of spare parts and tools should also be carried (see Table 11). Radios, fridges, and other

TABLE 11: Suggested equipment and supplies

General items

Appropriate fuel, water and food supplies, plus reserves
Jerry cans with an appropriate funnel or siphon
Maps
Car fridge
Camping gear
Clothing suitable for mild days and cold nights in winter months

Recovery items

Towing points at front and rear of vehicle
Long-handled shovel
Snatchum strap with DLI-approved shackles

Emergency items

Flying Doctor Radio
Fire extinguisher
Approved first aid kit

Essential tools

Assorted screwdrivers
Appropriate open-ended/ring spanners
Shifting spanner and plug spanner
Jumper leads
Pliers (long nosed and bull nosed)
Vice grips/multi grips
Socket set
G-clamp
Hammer and cold chisel
Hacksaw
Spring clamps

Basic camping equipment

Tent or swag
Ground sheets
Folding chairs and table
Lantern
Torch or fluorescent light
Length of cord for clothes line
Strong rubbish bags
Sleeping bag and pillow (optional)
Cooking equipment
Washing equipment for dishes and clothes

Vehicle spare parts

Radiator hoses (top and bottom) and clamps
Radiator coolant or distilled water
Fan belt
Auxiliary belts for air-conditioning, power steering and the like
Spark plugs and points
Fuel filter
Engine and transmission oils
Brake fluid
Workshop manual
Assorted nuts, bolts and washers
Range of spare fuses
Insulation/masking tape
Spare electrical wire
Spare plastic tubing or fuel line hose
Small tin of grease
Small coil of fencing wire
'WD40', 'CRC' or 'RP7'
Glues and/or sealants, such as epoxy-based 'Plastibond', 5-minute 'Araldite' and silicon-based 'Silastic'
Small electric soldering iron
Water pump, fuel pump and oil filter (for vehicles with mileage in excess of 50,000 km)
2 spare wheels
Quality 12-volt air compressor
Tyre gauge
At least 2 spare tubes, particularly if vehicle has tubeless tyres
Tyre/tube repair kit
Appropriate tyre-changing equipment, such as vehicle jack, solid 30-cm^2 base plate (for example a block of wood), wheelbrace, tyre levers, beadbreaker, rubber mallet

items which may have been stored for long periods should be checked that they are in sound working order. Just prior to departure, obtain current information on the condition of your intended route from an appropriate authority, such as the local police at Laverton or Coober Pedy, park rangers at Ceduna, Port Augusta or Kalgoorlie, or a relevant station owner. Advise authorities and friends of your departure date, route and estimated time of arrival at the intended destination. Consider making prescheduled radio contacts during the course of your journey. Confirm your safe arrival after reaching the destination.

Crossing the Desert

Most people travel along one of four main routes across the desert: the Anne Beadell Highway, the Vokes Hill Corner to Cook Road, the Connie Sue Highway, and the road between Laverton and Warburton.

Anne Beadell Highway

This major east–west route links Coober Pedy and Laverton via Mabel Creek Station, Tallaringa Conservation Park, Emu, the Maralinga Lands, Vokes Hill Corner in the Unnamed Conservation Park, the Serpentine Lakes and the South Australian–Western Australian border, Neale Junction and its adjoining Nature Reserve, and the Yeo Lake Nature Reserve. The approximate distances between major points of interest are:

	Distance (km)*
Coober Pedy to Mabel Creek Station	50
Mabel Creek to Tallaringa Well	110
Tallaringa Well to Emu	125
Emu to Vokes Hill Corner	155
Vokes Hill Corner to Serpentine Lakes	170
Serpentine Lakes to Neale Junction	340
Neale Junction to Laverton	375
Total distance	1325

*Please note these (and subsequent) distances listed in this Appendix are only accurate to within ±20 kilometres.

The following permits will be required to traverse this route.

- *Woomera Prohibited Area*
  Emu lies within the Woomera Prohibited Area. A permit to transit this area must be obtained from the Area Administrator, Defence Support Centre, PO Box 157, Woomera SA 5270, telephone (086) 743370 or fax (086) 743308.
- *Unnamed Conservation Park*
  Camping in the Unnamed Conservation Park is permitted within 100 metres of the Anne Beadell Highway (except for a 50-kilometre zone between longitudes 129°20'E and 129°48'E which has been specifically designated a no-camping area). A permit for overnight camping is required and may be obtained from the Department of Environment and Natural Resources (DENR) Far West District Office, PO Box 569, Ceduna SA 5690, telephone (086) 253144 or fax (086) 253123. A camping fee of $2 per person per night is also payable.
- *Maralinga Lands*
  The section of the Anne Beadell Highway west of the Woomera Prohibited Area and east of the Unnamed Conservation Park is within the Maralinga Lands. A permit to transit the Maralinga Lands must be obtained from Maralinga Tjarutja Inc., PO Box 435, Ceduna SA 5690, telephone (086) 252946 or fax (086) 253076.

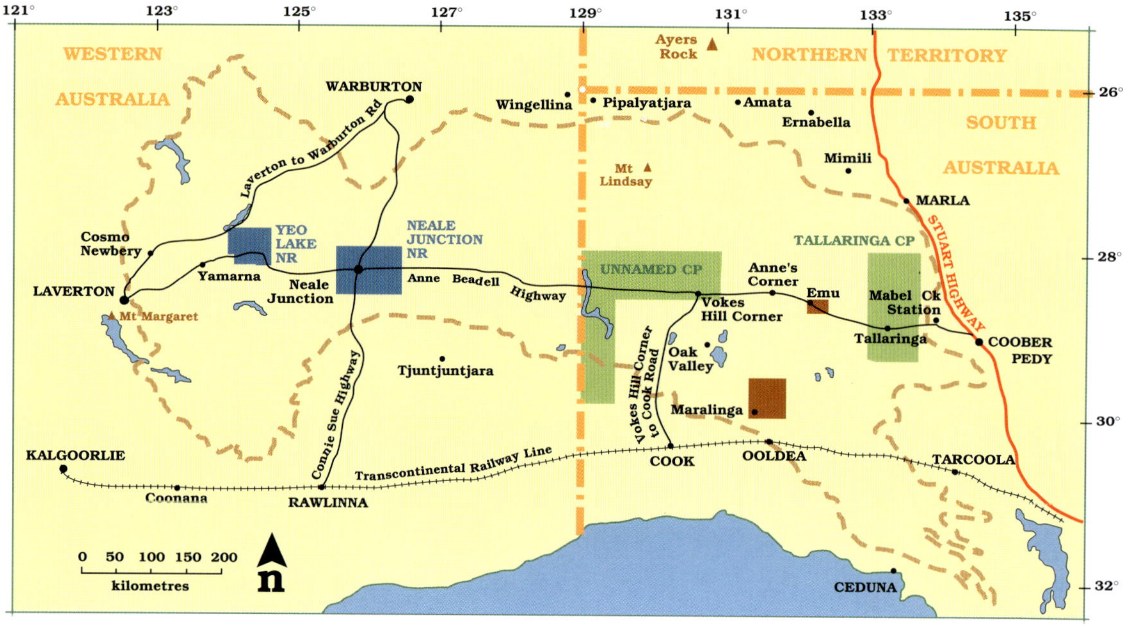

The main access roads across the Great Victoria Desert.

- *Mabel Creek Station*
  The Tjaliri Aboriginal Corporation has recently purchased Mabel Creek Station on the desert's eastern edge. New agreements for access through the station to the Anne Beadell Highway are currently being discussed by Tjaliri, the Manager of the North Region of the Department of Environment and Natural Resources (DENR) and the South Australian Pastoral Board.
  At present, the Station Manager of Mabel Creek should be advised of your intentions to pass through this station. The Manager may be contacted at Mabel Creek Station via Coober Pedy, SA 5723, telephone (086) 725204.
- *Western Australian Nature Reserves*
  There are no formal permit requirements for private access across the Neale Junction or Yeo Lake Nature Reserves (or any other A Class Nature Reserve in the Great Victoria Desert).

Vokes Hill Corner to Cook Road

The total distance of this road is approximately 280 kilometres. It is an approved access road, passing through both the Unnamed Conservation Park and the Maralinga Lands. The appropriate permits therefore need to be obtained from the DENR Far West District Office and Maralinga Tjarutja before traversing this road.

Please note: If you intend to travel on roads within the Maralinga Lands other than the Anne Beadell Highway or the Vokes Hill Corner to Cook Road, specific written approval must first be obtained from Maralinga Tjarutja.

Beryl Mazzachi

Mark Shephard

TOP: The visitor to the Australian outback will be captivated by the exquisite patterning and rich colour of the desert sands.
ABOVE: Sunset falls on the Anne Beadell Highway...

Connie Sue Highway

Approximate distances along this track are as follows:

	Distance (km)
Rawlinna to Neale Junction	335
Neale Junction to Warburton	315
Total distance	650

The northern part of the Connie Sue Highway passes through Ngaanyatjarra Lands. A transit permit must be obtained from the Ngaanyatjarra Council (Aboriginal Corporation), PO Box 644, Alice Springs, NT 0871, telephone (089) 534955 or fax (089) 531892.

Laverton to Warburton

The total distance of this road is approximately 560 kilometres. A transit permit must also be obtained from the Ngaanyatjarra Council to travel this route.

Please note: If you intend to travel on roads within the Ngaanyatjarra Lands other than the Connie Sue Highway or the Laverton to Warburton road, specific written approval must first be obtained from the Ngaanyatjarra Council.

Other Tracks Across the Desert

Non-Aboriginal visitation of specific Aboriginal access tracks, such as the Great South Road from Wingellina to Tjuntjuntjara and the road from Oak Valley to Tjuntjuntjara, is not permitted unless prior written approval has been granted by the appropriate Aboriginal Council(s).

Several other tracks pass through or near to Maralinga. This location is under the control of the Test Site Management Branch of the Commonwealth Department of Primary Industries and Energy in Canberra. Formal written permission must be obtained from this Department to travel through Maralinga. It is strongly advised that the Maralinga area be avoided.

Desert Etiquette

The visitor to the Great Victoria Desert has a responsibility to treat this unique and fascinating area with respect and to minimise his/her impact on its environment. Attempt to practise the following codes of bush etiquette:

- Remain on well-defined, approved tracks at all times. Reduce tyre pressure according to track conditions. This will enable smoother passage over sandy sections of desert roads and assist in preventing corrugation or scalloping of road surfaces. Consider other travellers who may be following your route! Do not drive on salt lakes as this practice may cause long-term scarring of the lake's surface. Do not cut corners or create new tracks around obstacles such as fallen trees. If entering the desert via pastoral stations, leave gates as you find them. If a gate is closed, then close it again after you have passed through.
- Camp in natural clearings, rather than in sand dunes where vegetation is more easily damaged. Avoid making unnecessary vehicle tracks around a campsite and take care not to drive over vegetation while positioning your vehicle for camp.

- Keep campfires small and attempt to conserve wood. Remember fallen timber that is hollow or cracked may provide a home for small reptiles, mammals and even birds. Never cut down branches from a living tree. Campfire remains should be as unobtrusive as possible. Do not burn plastic, tin or foil.
- Take out what you bring in! Place all rubbish in strong plastic bags and carry them out of the desert. Do not bury rubbish, as it disturbs the soil and is likely to be dug up or scattered by dingoes, birds and subsequently the wind. Burn any toilet paper used.
- Firearms are not permitted in parks or Aboriginal reserves. Keep domestic pets, particularly dogs and cats, at home.
- Enjoy, record and observe the desert's fauna quietly and without interfering with their habits or lifestyle. Do not pick or collect native flora, unless you have a permit specifically for this purpose.
- Do not disturb sites, objects or places of significance to Aboriginal people. Leave all relics and artefacts as you find them. Avoid the temptation to 'souvenir' such items.
- A significant part of a desert experience is the enjoyment of the sounds of the bush. Attempt to keep noise levels to a minimum and avoid the use of generators.

Bibliography

Barker WB and Greenslade PJM (eds) (1982). *Evolution of the flora and fauna of arid Australia*. Peacock Publications, Adelaide.

Bates D (1938). *The passing of the Aborigines*. John Murray, London.

Beadell L (1989). Six titles, namely *Too Long in the bush, Blast the bush, Bush bashers, Still in the bush, Beating about the bush,* and *End of an era.* Weldon Publishing, Sydney.

Beard JS (1990). *Plant Life of Western Australia*. Kangaroo Press, WA.

Berndt R and Berndt C (1981). *The world of the first Australians*. Revised edn. Lansdowne, Sydney.

Bolam AG (1923). *The Trans-Australian Wonderland*. Modern Printing, Melbourne.

Boomsma CD (1981). *Native trees of South Australia*. Bulletin 19 Second edn. Woods and Forests Department, SA.

Bryce S (ed.) (1992). *Women's gathering and hunting in the Pitjantjatjara homelands.* (Book and video available.) Institute for Aboriginal Development, Alice Springs.

Burbidge AA, McKenzie NL, Chapman A and Lambert PM (1976). The wildlife of some existing and proposed reserves in the Great Victoria and Gibson Deserts, Western Australia. *Wildlife Research Bulletin Western Australia* Vol. 5, pp. 1–16.

Cannon M (1987). *The exploration of Australia*. Readers Digest, Sydney.

Carnegie DW (1898). *Spinifex and sand.* Facsimile edn produced in 1989 by Hesperian Press, WA.

Cogger HG (1992). *Reptiles and amphibians of Australia*. Reed Books, Sydney.

Copley PB and Kemper CM (eds) (1992). *A biological survey of the Yellabinna region South Australia.* South Australian National Parks and Wildlife Service and Department of Environment and Land Management, Adelaide.

Davis PS and Kirke BK (1991). *Kapi Palya. Meeting the water requirements of an Aboriginal community at Oak Valley, Maralinga Lands, South Australia.* South Australian Health Commission, Adelaide.

Drexel JF, Preiss WV and Parker AJ (eds) (1993). *The geology of South Australia. Volume 1 The Pre Cambrian.* Bulletin 54. Department of Mines and Energy SA.

Duguid C (1972). *Doctor and the Aborigines*. Rigby, Adelaide.

Dutton G (1970). *Australia's last explorer: Ernest Giles*. Faber and Faber, London.

Edwards WH (1993). *An introduction to Aboriginal societies*. Social Science Press, New South Wales.

Far West Region, Department of Environment and Land Management (1993). *Unnamed Conservation Park Management Plan.* Department of Environment and Land Management, SA.

Flannery T, Kendell P and Wynn-Moylan K (1990). *Australia's vanishing mammals*. RD Press, Sydney.

Garnett S (ed.) (1993). *Threatened and Extinct Birds of Australia*. Second Edition. RAOU Report 82. Royal Australasian Ornithologists Union and Australian National Parks and Wildlife Service, Melbourne.

Giles E (1889). *Australia twice traversed.* Two vols. Paperback facsimile edn produced in 1986 by Doubleday, Sydney.

Glass A (1993). *Into another world. A glimpse of the culture of the Ngaanyatjarra people of Central Australia.* Institute for Aboriginal Development, Alice Springs.

Green N (1990). *Desert School.* Fremantle Arts Centre Press, WA.

Greenslade P, Joseph L and Barley R (eds) (1986). *The Great Victoria Desert.* Nature Conservation Society of South Australia, Adelaide.

Heatwole H and Taylor J (1987). *Ecology of reptiles*. Surrey Beatty and Sons, Sydney.

Hilliard W (1968). *The people in between.* Hodder and Stoughton, London.

Institute for Aboriginal Development (1988). *Puna. Yankunytjatjara plant use*. Angus and Robertson, Sydney.

Kean J, Richardson G and Trueman N (eds) (1988). *Aboriginal role in nature conservation*. Emu Conference June 7th–9th, 1988.

Laut P, Heyligers PC, Gael Keig et al. (1977). *Environments of South Australia Province 7 Western Pastoral.* CSIRO Division of Land Use Research, Canberra.

Laut P, Heyligers PC, Gael Keig et al. (1985). *Environments of South Australia Province 8 Northern Arid.* CSIRO Division of Land Use Research, Canberra.

Mattingley C and Hampton K (eds) (1992). *Survival in our own land.* Hodder and Stoughton, Sydney.

McClelland J, Fitch J and Jonas WJA (1985). *The report of the Royal Commission into British nuclear tests in Australia.* AGPS, Canberra.

McKenzie NL and Burbidge AA (eds) (1979). The wildlife of some existing and proposed Nature Reserves in the Gibson, Little Sandy and Great Victoria Deserts, Western Australia. *Wildlife Research Bulletin Western Australia* Vol. 8, pp. 1–36.

McKenzie NL and Robinson AC (eds) (1987). *A biological survey of the Nullarbor region, South and Western Australia in 1984.* South Australian Department of Environment and Planning, Western Australian Department of Conservation and Land Management and the Australian National Parks and Wildlife Service.

Milliken R (1986). *No conceivable injury.* Penguin Books, Melbourne.

Morelli J (1992). *Fire management in the Great Victoria Desert.* Australian National Parks and Wildlife Service and South Australian National Parks and Wildlife Service, Adelaide.

Morgan M (1986). *A drop in the bucket. The Mount Margaret Story.* United Aborigines Mission, Box Hill, Victoria.

Morton P (1989). *Fire across the desert.* AGPS, Canberra.

Peake-Jones K (1991). *Facsimile edition of the handbook of instructions for the Elder Exploration Expedition together with A Study of Incompatibles.* Royal Geographical Society of Australasia, South Australian Branch, Adelaide.

Salter E (1971). *Daisy Bates.* Coward, McCann and Geoghegan, New York.

Steele W and C (1978). *To the Great Gulf. The surveys and explorations of LA Wells.* Lynton Publications, Adelaide.

Strahan R (ed.) (1983). *The Australian Museum Complete Book of Australian Mammals.* Angus and Robertson, Sydney.

Symonds JL (1985). *A history of British atomic tests in Australia.* AGPS, Canberra.

Toyne P and McLaren N (1988). *Great Victoria Desert and Nullarbor Park World Heritage Listing.* A proposal to the South Australian Government from the Australian Conservation Foundation. Held by the library of the Conservation Council of South Australia, 120 Wakefield Street, Adelaide.

Toyne P and Vachon D (1987). *Growing up the country. The Pitjantjatjara struggle for their land.* Penguin Books, Melbourne.

Turner V (1950). *Ooldea.* S John Bacon, Melbourne.

Whibley DJE and Symon DE (1992). *Acacias of South Australia.* Government Printer, SA.

References

Introduction Setting the Record Straight

1. Beard JS (1969). The natural regions of the deserts of Western Australia. *Journal of Ecology* Vol. 57, pp. 677–706.
2. Beard JS (1990). *Plant Life of Western Australia.* Kangaroo Press, WA.
3. Giles E (1889). *Australia twice traversed.* Two vols. Paperback facsimile edn produced in 1986 by Doubleday, Sydney.
4. Toyne P and McLaren N (1988). *Great Victoria Desert and Nullarbor Park World Heritage Listing.* A proposal to the South Australian Government from the Australian Conservation Foundation. Held by the library of the Conservation Council of South Australia, 120 Wakefield Street, Adelaide.

Chapter 1 Biophysical Description

1. Morelli J (1992). *Fire management in the Great Victoria Desert.* Australian National Parks and Wildlife Service and South Australian National Parks and Wildlife Service, Adelaide.
2. Benbow MC (1992). Geomorphology and geology. In Copley PB and Kemper CM (eds), *A biological survey of the Yellabinna region South Australia.* South Australian National Parks and Wildlife Service and Department of Environment and Land Management, Adelaide, pp. 15–38.
3. Davis PS and Kirke BK (1991). *Kapi Palya. Meeting the water requirements of an Aboriginal community at Oak Valley, Maralinga Lands, South Australia.* South Australian Health Commission, Adelaide.
4. Beard JS (1990). *Plant Life of Western Australia.* Kangaroo Press, WA.
5. Lowry DC (1970). Geology of the Western Australian part of the Eucla Basin. *Western Australian Geological Survey,* Bulletin 122.
6. Drexel JF, Preiss WV and Parker AJ (eds) (1993). *The geology of South Australia. Vol. 1. The Pre Cambrian,* Bulletin 54, Department of Mines and Energy SA.
7. McKenzie NL and Robinson AC (eds) (1987). *A biological survey of the Nullarbor Region South and Western Australia in 1984.* South Australian Department of Environment and Planning, Western Australian Department of Conservation and Land Management and the Australian National Parks and Wildlife Service.
8. Benbow MC (1990). Tertiary coastal dunes of the Eucla Basin, Australia. *Geomorphology* Vol. 3, pp. 9–29.
9. Langford-Smith T (ed.) (1977). *Silcretes in Australia.* New England University Press, Armidale, NSW.
10. Laut P, Heyligers PC, Gael Keig et al. (1977). *Environments of South Australia Province 7 Western Pastoral.* CSIRO Division of Land Use Research, Canberra.
11. Laut P, Heyligers PC, Gael Keig et al. (1985). *Environments of South Australia Province 8 Northern Arid.* CSIRO Division of Land Use Research, Canberra.
12. Fotheringham D (1986). Soils and landforms. In Greenslade P, Joseph L and Barley R (eds),*The Great Victoria Desert.* Nature Conservation Society of South Australia, Adelaide, pp. 3–12.
13. Lowry DC and Jennings JN (1974). The Nullarbor karst, Australia. *Zeitschrift fur Geomorphologie* NF Vol. 18, pp. 35–81.

Chapter 2 A Veritable Botanic Garden

1. Beard JS (1990). The Eremean Province (Ch. 4) in *Plant Life of Western Australia.* Kangaroo Press, WA, pp. 178–185.
2. Pearson DJ (1994). The vegetation and flora of Queen Victoria Spring Nature Reserve, Western Australia. Unpublished MSc thesis. Department of Botany, University of Western Australia.
3. Burbidge AA, McKenzie NL, Chapman A and Lambert PM (1976). The wildlife of some existing and proposed reserves in the Great Victoria and Gibson Deserts, Western Australia. *Wildlife Research Bulletin Western Australia* Vol. 5, pp. 1–16.
4. McKenzie NL and Burbidge AA (eds) (1979). The wildlife of some existing and proposed Nature Reserves in the Gibson, Little Sandy and Great Victoria Deserts, Western Australia. *Wildlife Research Bulletin Western Australia* Vol. 8, pp. 1–36.
5. Beard JS (1974). The vegetation of the Great Victoria Desert area. Explanatory notes to sheet 3 of Vegetation Survey of Western Australia: Great Victoria Desert. University of WA Press, Nedlands, WA.
6. Vickery J (1967). *Far West and Far North-West interior. South Australian Pastoral Board inspection.* 19 June to 19 July 1967. Internal report held by the Pastoral Board.
7. Douglas JS, Williams L, Levitzke V and Taverner B (1979). *Survey of the Unnamed Conservation Park using Landsat.* Ecological Survey Unit, South Australian Department for the Environment, Adelaide.
8. Williams L (1986). Vegetation of the eastern Great Victoria Desert: structural analysis. In Greenslade P, Joseph L and Barley R (eds), *The Great Victoria Desert.* Nature Conservation Society of South Australia, Adelaide, pp. 24–32.
9. Symon DE and Copley PB (1986). Vegetation of the eastern Great Victoria Desert: floristics. In Greenslade P, Joseph L and Barley R (eds), *The Great Victoria Desert.* Nature Conservation Society of South Australia, Adelaide, pp. 14–24.
10. Davies R, Copley P and Pearman C (1986). Vegetation of the eastern Great Victoria Desert: patterns of distribution. In Greenslade P, Joseph L and Barley R (eds), *The Great Victoria Desert.* Nature Conservation Society of South Australia, Adelaide, pp. 33–49.

11. Morelli J (1992). *Fire management in the Great Victoria Desert*. Australian National Parks and Wildlife Service and the South Australian National Parks and Wildlife Service.
12. Robinson AC, Copley PB and Drewien GN (1988). *A floristic vegetation map of the Tallaringa area*. National Parks and Wildlife Service and the Department of Environment and Planning, SA.
13. Lay BG and Magery A (1985). A botanical survey of P.E.L. 24 Lake Phillipson area, South Australia. Unpublished report for CRA Ltd.
14. Johnson A and Venning J (1982). *Wilkinson Lakes Region Environmental Study. Environmental implications of development for pastoral use*. Department of Environment and Planning, SA.
15. Robinson AC and Canty PD (1992). Vegetation. In Copley PB and Kemper CM (eds), *A biological survey of the Yellabinna region South Australia*. National Parks and Wildlife Service and Department of Environment and Land Management, SA, pp. 69–120.
16. Beard JS (1969). The natural regions of the deserts of Western Australia. *Journal of Ecology* Vol. 57, pp. 677–711.
17. Brooker MH and Kleinig DA (1990). *Field guide to Eucalypts. Volume 2*. Inkata Press, Melbourne.
18. Boomsma CD (1981). *Native trees of South Australia*. Bulletin 19, 2nd edn. Woods and Forests Department, SA.
19. Brooker MIH and Hopper SD (1993). Eucalyptus. *Nuytsia* Vol. 9, pp. 32–39 (Journal of the Western Australian Herbarium).
20. Boomsma CD (1988). A new *Eucalyptus* species from the Wyola region, in the far west of South Australia. *Journal of the Adelaide Botanical Gardens* Vol. 10, pp. 59–61.
21. Whibley DJE and Symon DE (1992). *Acacias of South Australia*. Government Printer, SA.
22. Jessop J (ed.) (1981). *Flora of Central Australia*. The Australian Systematic Botany Society. Reed Books Pty Ltd, Sydney.
23. Urban A (1990). *Wildflowers of Central Australia*. Southland Editions, Port Melbourne, Victoria.

Chapter 3 BIRDS

1. Cleland JB (1942). Further notes on early ornithology in South and Central Australia. Notes on birds in the journal of exploration by RT Maurice in 1901 and 1902. *South Australian Ornithologist* Vol. 16, pp. 41–43.
2. Cleland JB (1946). Reference to geographic travel in central Australia 1872–4 by Ernest Giles. *South Australian Ornithologist* Vol. 18, p. 6.
3. Cleland JB (1937). The history of ornithology in South Australia. *South Australian Ornithologist* Vol. 37, pp. 33–45.
4. White SA (1919). Four ornithological trips to the Nullarbor Plain. *Emu* Vol. 18, pp. 189–198.
5. Cleland JB (1923). Notes on the birds of Ooldea. *South Australian Ornithologist* Vol. 7, pp. 17–21.
6. McGilp JN (1932). Birds of the Nullarbor Plain and far west coast of South Australia. *South Australian Ornithologist* Vol. 1, pp. 146–151.
7. Le Souef AS (1921). Notes on birds observed at Ebor and the Nullarbor Plain. *Emu* Vol. 21, pp.125–128.
8. Whitlock FL (1922). Notes from the Nullarbor Plain. *Emu* Vol. 21, pp. 170–187.
9. Bolam AG (1923). *The Trans-Australian Wonderland*. Modern Printing, Melbourne.
10. Slater P and Lindgren E (1957). A visit to Queen Victoria Spring, January 1955. *Western Australian Naturalist* Vol. 5, pp. 10–18.
11. Serventy V (1957). Birds of Queen Victoria Spring. *Western Australian Naturalist* Vol. 5, pp. 102–107.
12. Ford J and Sedgwick EH (1967). Bird distribution in the Nullarbor Plain and Great Victoria Desert region, Western Australia. *Emu* Vol. 67, pp. 99–124.
13. Ford J (1971). Distribution and taxonomy of southern birds in the Great Victoria Desert. *Emu* Vol. 71, pp. 27–36.
14. Brooker MG, Ridpath MG, Estbergs AJ et al. (1979). Bird observations on the north-western Nullarbor Plain and neighbouring regions, 1967–1978. *Emu* Vol. 79, pp. 176–190.
15. Burbidge AA, McKenzie NL, Chapman A and Lambert PM (1976). The wildlife of some existing and proposed reserves in the Great Victoria and Gibson Deserts, Western Australia. *Wildlife Research Bulletin Western Australia* Vol. 5, pp. 1–16.
16. McKenzie NL and Burbidge AA (eds) (1979). The wildlife of some existing and proposed nature reserves in the Gibson, Little Sandy and Great Sandy Deserts, Western Australia. *Wildlife Research Bulletin Western Australia* Vol. 8, pp. 1–36.
17. Burbidge AA, Casperson KD and Fuller PJ (1987). Birds. In McKenzie NL and Robinson AC (eds), *A biological survey of the Nullarbor region South and Western Australia in 1984*. South Australian Department of Environment and Planning, Western Australian Department of Conservation and Land Management and the Australian National Parks and Wildlife Service, pp. 153–178.
18. Hunt T (1993). A wet year in the Great Victoria Desert. *Wingspan* March issue, pp. 16–18.
19. Black AB and Badman FJ (1986). Birds of the eastern Great Victoria Desert. In Greenslade P, Joseph L and Barley R (eds), *The Great Victoria Desert*. Nature Conservation Society of South Australia, Adelaide, pp. 66–94.
20. Matthew J and Carpenter G (1990). *A survey of birds and their habitats of the Unnamed Conservation Park and adjacent eastern Great Victoria Desert*. Report for the South Australian Ornithological Association (held in the Association's library).
21. Close DH and Jaensch RP (1984). Birds of the north-west of South Australia. *South Australian Ornithologist* Vol. 29, pp. 81–99.
22. Shephard M (1993). *A bird survey of the Pitjantjatjara Lands*. An internal report prepared for the Pitjantjatjara Council, August 1993.
23. Shephard M (1994). *Birds of the western Pitjantjatjara Lands*. An internal report prepared for the Pitjantjatjara Council, August 1994.
24. Cohen B, Reid J and Casperson K (1992). Birds. In Copley PB and Kemper CM (eds), *A biological survey of the Yellabinna region South Australia*. National Parks and Wildlife Service of South Australia and the Department of Environment and Land Management, pp. 137–162.

25. Crossley L (1994). *Expedition Tallaringa*. Preliminary report. Australian and New Zealand Scientific Exploration Society, Melbourne.
26. Schodde R (1982). Origin, adaptation and evolution of birds in arid Australia. In Barker WB and Greenslade PJM (eds), *Evolution of the flora and fauna of arid Australia*. Peacock Publications, Adelaide, pp. 191–224.
27. Ford J (1974). Speciation in Australian birds adapted to arid habitats. *Emu* 74, 161–168.
28. Ford J (1987). Minor isolates and minor geographical barriers in avian speciation in continental Australia. *Emu* Vol. 87, pp. 90–102.
29. Ford J (1987). Hybrid zones in Australian birds. *Emu* Vol. 87, pp. 158–178.
30. Garnett S (ed.) (1993). *Threatened and Extinct Birds of Australia*. Second Edition. RAOU Report 82. Royal Australasian Ornithologists Union and Australian National Parks and Wildlife Service, Melbourne.
31. Robinson AC, Casperson KD and Copley PB (1990). Breeding records of malleefowl *Leiopa ocellata* and scarlet-chested parrots *Neophema splendida* within the Yellabinna Wilderness Area, South Australia. *South Australian Ornithologist* Vol. 31, pp. 8–12.
32. Chapman A (CALM, Kalgoorlie). Personal communication.
33. Robinson AC, Copley PB and Drewein GN (1988). *A floristic vegetation map of the Tallaringa area*. National Parks and Wildlife Service and the Department of Environment and Planning, SA.
34. Black A. Personal communication.
35. Carpenter G. Personal communication.
36. Henderson RC (1977). Breeding records of the scarlet-chested parrot in the Great Victoria Desert, and further records in the lower north-east. *South Australian Ornithologist* Vol. 27, pp. 222–223.
37. Joseph L (1976). Recent records of the scarlet-chested parrot. *South Australian Ornithologist* Vol. 27, pp. 144–145.
38. Stewart DA and Shephard MDS (1995). Large flocks of scarlet-chested parrots in the Great Victoria Desert. *South Australian Ornithologist* (in press).
39. Andrew D and Palliser T (1993). Observations of scarlet-chested parrots. *Wingspan* June issue, pp. 4–5.
40. Allen R (1987). Alexandra's parrots in the Great Victoria Desert. *South Australian Ornithologist* Vol. 30, p. 75.
41. Aspley R. Personal communication.
42. Carter M (1993). Alexandra's or Princess Parrot. *Wingspan* Dec. issue, pp. 32–35.
43. Fleming M (Conservation Commission of the Northern Territory, Alice Springs). Personal communication.

Chapter 4 MAMMALS

1. Gara T (1993). Early observations of the mammals of the Nullarbor Plain and Great Victoria Desert in South Australia. Unpublished manuscript. Copy held by the South Australian Museum.
2. Boscacci LJ, McKenzie NL and Kemper CM (1987). Mammals. In McKenzie NL and Robinson AC (eds), *A biological survey of the Nullarbor region South and Western Australia 1984*. South Australian Department of Environment and Planning, Western Australian Department of Conservation and Land Management and Australian National Parks and Wildlife Service, pp. 103–152.
3. Bolam AG (1923). *The Trans-Australian Wonderland*. Modern Printing, Melbourne.
4. Kemper CM (1992). Mammals. In Copley PB and Kemper CM (eds), *A biological survey of the Yellabinna region South Australia*. National Parks and Wildlife Service and the Department of Environment and Land Management, Adelaide, pp. 121–136.
5. Burbidge AA, McKenzie NL, Chapman A and Lambert PM (1976). The wildlife of some existing and proposed reserves in the Great Victoria and Gibson Deserts, Western Australia. *Wildlife Research Bulletin of Western Australia* Vol. 5, pp. 1–16.
6. McKenzie NL and Burbidge AA (eds) (1979). The wildlife of some existing and proposed nature reserves in the Gibson, Little Sandy and Great Victoria Deserts, Western Australia. *Wildlife Research Bulletin of Western Australia* Vol. 8, pp. 1–36.
7. Barritt MK (1986). Mammals. In Greenslade P, Joseph L and Barley R (eds), *The Great Victoria Desert*. Nature Conservation Society of South Australia, pp. 113–118.
8. Robinson AC, Copley PB and Drewien GN (1988). Appendix II. The mammals recorded in the Tallaringa study area. In *A floristic vegetation map of the Tallaringa area*. National Parks and Wildlife Service and the Department of Environment and Planning, Adelaide.
9. Crossley L (1994). *Expedition Tallaringa*. Preliminary report. Australian and New Zealand Scientific Exploration Society, Melbourne.
10. Morrison RGB (1981). *A field guide to the tracks and traces of Australian animals*. Rigby Publishers, Adelaide.
11. Burbidge AA, Johnson KA, Fuller PJ and Southgate RI (1988). Aboriginal knowledge of the mammals of the central deserts of Australia. *Australian Wildlife Research* Vol. 15, pp. 9–39.
12. Pearson D (1991). First records of the mulgara, *Dasycercus cristicauda*, from the Gibson Desert and Queen Victoria Spring Nature Reserves. *Western Australian Naturalist* Vol. 18, pp. 159–161.
13. Hart RP and Kitchener DJ (1986). First record of *Sminthopsis psammophila* (Marsupialia: Dasyuridae) from Western Australia. *Records of the Western Australian Museum* Vol. 13, pp. 139–144.
14. Pearson D and Robinson AC (1990). New records of the sandhill dunnart, *Sminthopsis psammophila* (Marsupialia: Dasyuridae) in South and Western Australia. *Australian Mammology* Vol. 13, pp. 57–59.
15. Strahan R (ed.) (1983). *The Australian Museum Complete Book of Australian Mammals*. Angus and Robertson, Sydney.
16 Chapman A (CALM, Kalgoorlie). Personal communication.
17. Copley P (Biological Conservation Branch, DENR). Personal communication.
18. Burbidge A, Fuller P, Johnson K and Southgate R (1987). Vanishing desert mammals. *Landscope* Vol. 2, pp. 7–12.

19. Copley PB, Kemper CM and Medlin GC (1989). The mammals of north-western South Australia. *Records of the South Australian Museum* Vol. 23, pp. 75–88.
20. Finlayson HH (1961). On central Australian mammals. Part IV. The distribution and status of central Australian species. *Records of the South Australian Museum* Vol. 14, pp. 141–191.
21. Maurice RT (1897–1903). Maurice's manuscripts are held by both the Mortlock Library of South Australiana and the library of the South Australian Branch of the Royal Geographical Society of Australasia, Adelaide.
22. Stirling EC and Zietz A (1893). Scientific results of the Elder Exploring Expedition, Vertebrata. *Transactions of the Royal Society of South Australia* Vol. 16, pp. 154–176.
23. Seebeck JH, Brown PR, Wallis RL and Kemper CM (eds) (1989). *Bandicoots and bilbies*. Surrey Beatty and Sons in association with the Australian Mammal Society, Sydney.
24. Basedow H (1915). Journal of the Government North-West Expedition (March 30 – November 5, 1903). *Proceedings of the Royal Geographical Society of South Australia* Vol. 15, pp. 57–242.
25. Lindsay D (1893). *Journal of the Elder Scientific Expedition, 1891*. South Australian Parliamentary Papers 45, Adelaide.
26. Helms R (1896). Anthropology. In 'The scientific results of the Elder Exploring Expedition'. *Transactions of the Royal Society of South Australia* Vol. 16, pp. 237–332.
27. Dickman CR (1993). Raiders of the last ark: cats in inland Australia. *Australian Natural History* Vol. 24 (Winter issue), pp. 45–52.
28. Greenwood L (1993). CATastrophe. *Habitat Australia* May issue, pp. 12–14 .
29. Morton S (1989). The impact of European settlement on the vertebrate animals of Australia: a conceptual model. *Proceedings of the Ecological Society of Australia* Vol. 16, pp. 201–213.
30. Burbidge AA and McKenzie NL (1989). Patterns in the modern decline of Western Australia's vertebrate fauna: causes and implications. *Biological Conservation* Vol. 50, pp. 143–198.
31. Morelli J (1992). *Fire management in the Great Victoria Desert*. Australian National Parks and Wildlife Service and South Australian National Parks and Wildlife Service.

Chapter 5 REPTILES

1. Bird P (1986). Reptiles and amphibians of the eastern Great Victoria Desert. In Greenslade P, Joseph L and Barley R (eds), *The Great Victoria Desert*. Nature Conservation Society of South Australia, Adelaide, pp. 95–111.
2. Pianka ER (1969). Habitat specificity, speciation and species density in Australian desert lizards. *Ecology* Vol. 50, pp. 498–502.
3. Pianka ER (1969). Sympatry of desert lizards (*Ctenotus*) in Western Australia. *Ecology* Vol. 50, pp. 1012–1030.
4. Pianka ER (1992). A land of lizards. *Landscope* Autumn issue, pp. 11–16 .
5. Armstrong GP (1992). Reptiles. In Copley PB and Kemper CM (eds), *A biological survey of the Yellabinna region, South Australia*. National Parks and Wildlife Service and Department of Environment and Land Management, Adelaide, pp. 163–179.
6. Kemper CM, Reid J and Edwards A (1985). A survey of the vertebrate fauna of Mabel Creek (PEL 24). Unpublished report for CRA Ltd.
7. Robinson AC, Copley PB and Drewien GN (1988). Appendix IV. The amphibians and reptiles recorded in the Tallaringa study area to 1988. In *A floristic vegetation map of the Tallaringa area*. National Parks and Wildlife Service and the Department of Environment and Land Management, Adelaide.
8. Crossley L (1994). *Expedition Tallaringa*. Preliminary report. Australian and New Zealand Scientific Exploration Society, Melbourne.
9. McKenzie NL, Rolfe JK and Carter D (1987). Reptiles. In McKenzie NL and Robinson AC (eds), *A biological survey of the Nullarbor region South and Western Australia in 1984*. South Australian Department of Environment and Planning, Western Australian Department of Conservation and Land Management and the Australian National Parks and Wildlife Service, pp. 179–210.
10. Hutchinson M (Curator of Reptiles, South Australian Museum). Personal communication.
11. Cogger HG (1992). *Reptiles and amphibians of Australia*. Reed Books, Sydney.
12. Swanson S (1987). *Lizards of Australia*. Angus and Robertson, Sydney.
13. Cogger HG (1989). *Australian reptiles in colour*. Reed Books, Sydney.
14. Heatwole H and Taylor J (1987). *Ecology of reptiles*. Surrey Beatty and Sons, Sydney.
15. Pianka ER and Pianka HD (1976). Comparative ecology of twelve species of nocturnal lizards (Gekkonidae) in the Western Australian desert. *Copeia* (1), pp. 125–142.
16. Pianka ER and Pianka HD (1970). The ecology of *Moloch horridus* (Lacertilia: Agamidae) in Western Australia. *Copeia* (1), pp. 90–103.
17. Houston TF (1978). *Dragon lizards and goannas of South Australia*. South Australian Museum, Adelaide.
18. Pianka ER (1982). Observations on the ecology of *Varanus* in the Great Victoria Desert. *Western Australian Naturalist* Vol.15, pp. 37–44.
19. Burbidge AA, McKenzie NL, Chapman A and Lambert PM (1976). The wildlife of some existing and proposed reserves in the Great Victoria and Gibson Deserts, Western Australia. *Wildlife Research Bulletin of Western Australia* Vol. 5, pp. 1–16.
20. Pianka ER (1968). Notes on the biology of *Varanus eremius*. *Western Australian Naturalist* Vol. 11, pp. 29–44.
21. Pianka ER (1971). Notes on the biology of *Varanus tristis*. *Western Australian Naturalist* Vol. 11, pp. 180–183.
22. Schwaner T (1985). *Snakes in South Australia: a species list and overview*. South Australian Museum, Adelaide.
23. Mirtschin P and Davis R (1991). *Dangerous Snakes of Australia*. Ure Smith Press, Sydney.

Chapter 6 TRADITIONAL ABORIGINAL LIFE PRIOR TO EUROPEAN CONTACT

1. Smith MA (1989). The case for a resident human population in the Central Australian Ranges during full glacial aridity. *Archaeology in Oceania* Vol. 24, pp. 93–105.
2. Cane S (1990). *A proposal for joint management of the Unnamed Conservation Park. A report to Maralinga Tjarutja*. National Heritage Studies, ACT.
3. Berndt RM (1959). The concept of the "the tribe" in the Western Desert of Australia. *Oceania* Vol. 30, pp. 82–107.
4. Cane S (1991). *The spinifex people. Report to Paupiyala Tjarutja and ATSIC, Kalgoorlie, WA.* National Heritage Studies, ACT.
5. Gara T, Colley S, Brockwell S and Cane S (1988). *The Aboriginal metropolis of Ooldea*. An internal report prepared for the Aboriginal Heritage Branch, Department of Environment and Planning, South Australia. ANU Tech, Canberra.
6. Berndt R and C (1942). A preliminary report of fieldwork in the Ooldea region, western South Australia. *Oceania* Vol. 12, pp. 305–330.
7. Tindale N (1974). *Aboriginal tribes of Australia, their terrain, environmental controls, limits and proper names*. Australian National University Press, Canberra.
8. Brady M (1986). Leaving the spinifex: the impact of rations, missions and the atomic tests on the southern Pitjantjatjara. *Records of the South Australian Museum* Vol. 20, pp. 35–45.
9. Palmer K (1982). *Grey earth and clean sand. The Yalata Aborigines and their homelands.* Western Desert Project. Department of Psychiatry, School of Medicine, Flinders University of South Australia, Adelaide.
10. Davis PS and Kirke BK (1991). *Kapi Palya. Meeting the water requirements of an Aboriginal Community at Oak Valley, Maralinga Lands, South Australia*. South Australian Health Commission, Adelaide.
11. Berndt R (1941). Tribal migrations and myths centring on Ooldea, South Australia. *Oceania* Vol. 12, pp. 1–20.
12. Johnston TH (1941). Some Aboriginal routes in the western portion of South Australia. *Proceedings of the Royal Geographical Society of South Australia* Vol. 42, pp. 33–65.
13. Berndt R and C (1981). *The world of the first Australians* (rev. edn). Lansdowne, Sydney.
14. Edwards WH (1993). *An introduction to Aboriginal societies*. Social Science Press, Sydney.
15. Toyne P and Vachon D (1987). *Growing up the country*. Penguin Books, Melbourne.
16. Gara T (1989). Aborigines of the Great Victoria Desert: the ethnographic observations of the explorer Richard Maurice. *Journal of the Anthropological Society of South Australia* Vol. 27, pp. 15–47.
17. Morelli J (1992). *Fire Management in the Great Victoria Desert*. Australian National Parks and Wildlife Service and South Australian National Parks and Wildlife Service.
18. Cleland JB and Johnston TH (1942). Aboriginal names and uses of plants in the Ooldea region, South Australia. *Transactions of the Royal Society of South Australia* Vol. 66, pp. 93–103.
19. Maurice RT (1905). Across the great thirst land. *World Wide Magazine* Vol. 16, pp. 357–364.
20. Gara T (1995). *History of the Unnamed Conservation Park*. In preparation.
21. Brockwell S, Gara T, Colley S and Cane S (1989). The history and archaeology of Ooldea Soak and Mission. *Australian Archaeology* Vol. 28, pp. 55–77.
22. Bates DM (1938). *The passing of the Aborigines*. John Murray, Sydney.
23. Maurice RT. Papers relating to his expeditions (Accession no 158/1–56). Held by the Mortlock Library of South Australiana, Adelaide.
24. Bates DM (1921). Ooldea water. *Proceedings of the Royal Geographical Society of Australasia* (South Australian Branch) Vol. 21, pp. 73–78.
25. Bolam AG (1923). *Trans-Australian wonderland*. Modern Printing, Melbourne.
26. Berndt R and C (1942). Social organisation and institutions. *Oceania* Vol. 14, pp. 143–169.
27. Murray WR (1904). Explorations by RT Maurice—*"Fowlers Bay to Rawlinson Ranges" and "Fowlers Bay to Cambridge Gulf"*. South Australian Parliamentary Papers 43, Adelaide.
28. Bagshaw GC (1989). *Preliminary report on Aboriginal cultural interests in the Tallaringa Well region*. Aboriginal Heritage Branch, Department of Environment and Planning, Adelaide.
29. Crossley J (1994). *Expedition Tallaringa*. Preliminary report. Australian and New Zealand Scientific Exploration Society, Melbourne.
30. Talbot HWB and Clarke E deC (1917). *A geological reconnaissance of the country between Laverton and the South Australian border, including part of the Mount Margaret Goldfield.* Bulletin No. 75. Western Australia Geological Survey, Perth.
31. Giles E (1889). *Australia twice traversed*. Two vols. Paperback facsimile edn produced in 1986 by Doubleday, Sydney.
32. Carnegie DW (1898). *Spinifex and sand*. Facsimile edn produced in 1989 by Hesperian Press, WA.
33. Edwards WH (1983). Plants in Pitjantjatjara life and mythology. Lecture presented at the Society for Growing Australian Plants' Biennial Federal Conference and Seminar, 27 August – 9 September 1983, Adelaide.
34. Institute for Aboriginal Development (1988). *Puna. Yankunytjatjara plant use*. Angus and Robertson, Sydney.
35. Bryce S (ed.). *Women's gathering and hunting in the Pitjantjatjara homelands* (book and video available). Institute for Aboriginal Development, Alice Springs.
36. Hetzel BS and Frith HJ (1978). *The nutrition of Aborigines in relation to the ecosystem of Central Australia*. CSIRO, Melbourne.
37. Low T (1992). *Bush tucker*. Angus and Robertson, Sydney.
38. Isaacs J (1992). *Bush food*. Ure Smith Press, Sydney.

Chapter 7 SEVENY-FIVE YEARS OF EUROPEAN EXPLORATION (1857–1932)

1. Copley PB (1992). Land-use history. In Copley PB and Kemper CM (eds), *A biological survey of the Yellabinna region South Australia*. National Parks and Wildlife Service and Department of Environment and Land Management, Adelaide, pp. 38–53.
2. Hack S (1857–1858). *Explorations by Mr S. Hack.* South Australian Parliamentary Proceedings and Papers 156, Adelaide, pp. 2–11.

3. Warburton PE (1858). *Northern exploration. Reports, etc., of explorations into the interior by Messrs. Babbage, Warburton, Gerharty and Parry.* South Australian Parliamentary Proceedings and Papers 25, Adelaide, pp. 38–42.
4. Gerharty J (1858). *Northern exploration. Reports, etc., of explorations into the interior, by Messrs. Babbage, Warburton, Gerharty, and Parry.* South Australian Parliamentary Proceedings and Papers 25, Adelaide, pp. 42–47.
5. Hardman W (1865). *Explorations in Australia. The Journals of John McDouall Stuart.* Facsimile edn produced in 1984 by Hesperian Press, WA.
6. De Lisser EA (1865–1866). Journey into the interior of Australia, north-west of the Great Bight. *Proceedings of the Royal Geographical Society* (London) Vol. 10, pp. 129–130.
7. Gara T, Colley S, Brockwell S and Cane S (1988). *The Aboriginal metropolis of Ooldea.* A report to the Aboriginal Heritage Branch, South Australian Department of Environment and Planning. ANUTech, Canberra.
8. Ross J (1875). *Mr J Ross's explorations, 1874.* South Australian Parliamentary Papers 67, Adelaide.
9. Dutton G (1970). *Australia's last explorer: Ernest Giles.* Faber and Faber, London.
10. Feekin EHJ, Feekin GEE and Spate OHK (1970). *The discovery and exploration of Australia.* Thomas Nelson, Sydney.
11. Giles E (1889). *Australia twice traversed.* Two vols. Paperback facsimile edn produced in 1986 by Doubleday, Sydney.
12. Cannon M (1987). *The exploration of Australia.* Readers Digest, Sydney.
13. Tietkins WH (1879). *Diary of the exploration in South Australia of WH Tietkins 1879.* Typescript copy edited by EL Cook, Department of Supply, Weapons Research Establishment, SA.
14. Tietkins WH (1887). The Nullarbor Plains and the west boundary of the province. *Proceedings of the Royal Geographical Society of Australasia, South Australian Branch* Vol. 2, pp. 34–38.
15. Gara T (1995). History of the Unnamed Conservation Park. In preparation.
16. Jones JW (1880). *Examination of country north-east of Eucla.* South Australian Parliamentary Papers 191, Adelaide.
17. Chambers CJ (1884). Exploration from Arlcoodaby to Warburton Ranges. *The Adelaide Observer*, Dec. 27, p. 30.
18. Johnston TH (1941). Some Aboriginal routes in the western portion of South Australia. *Proceedings of the Royal Geographical Society of South Australia* Vol. 42, pp. 33–65.
19. Steele W and C (1978). *To the Great Gulf. The surveys and explorations of LA Wells.* Lynton Publications, Adelaide.
20. Peake-Jones K (1991). *Handbook of instructions for the Elder Exploration Expedition together with A Study of Incompatibles.* Facsimile edn produced by the Royal Geographical Society of Australasia, South Australian Branch, Adelaide.
21. Peake-Jones K (1985). The Elder Scientific Exploration Expedition, 1891: a study of incompatibles. *Proceedings of the Royal Geographical Society of Australasia, South Australian Branch* Vol. 85, pp. 54–67.
22. Shephard M (1994). *The Simpson Desert — natural history and human endeavour.* Reed Books, Sydney.
23. Lindsay D (1893). *Journal of the Elder Scientific Exploring Expedition, 1891.* South Australian Parliamentary Papers 45, Adelaide.
24. Coxon S. Personal communication to C Harris, Director of the Environment Division, Department of Environment and Planning, Adelaide, 5 March 1989.
25. Streich JV (1992). In search of the Elder Expedition 1990. *South Australian Geographical Journal* Vol. 92, pp. 68–77.
26. Macdonald B (1994). *Filling in the gaps. A cameo of the Royal Geographical Society of Australasia.* Supplement to the 1995 South Australian Year Book. ABS, Adelaide.
27. Carnegie DW (1898). *Spinifex and sand.* Facsimile edn produced in 1989 by Hesperian Press, WA.
28. Pearson DJ (1994). The vegetation and flora of Queen Victoria Spring Nature Reserve, Western Australia. Unpublished MSc thesis, Department of Botany, University of Western Australia, WA.
29. *Australian Dictionary of Biography*, Vol. 3 (1851–1890) A–C (entry on William Henry James Carr-Boyd). Melbourne University Press, Melbourne, 1969, pp. 357–358 .
30. Carr-Boyd W (1897). Journey from Western Australia to Warina, in South Australia. *Geographical Journal* Vol. 9, pp. 61–62.
31. Simpson M (1992). Samuel Grau Hubbe: a minor explorer. *South Australian Geographical Journal* Vol. 92, pp. 37–48.
32. Hubbe S (1897). *Stock route expedition from South to Western Australia.* South Australian Parliamentary Papers 51, Adelaide.
33. Mason A (1896). *Report of an expedition in the south-eastern portion of Western Australia to inquire into a reported incursion of rabbits.* Government Printer, Perth.
34. Brown HYL (1898). *Government geologist's report on exploration in western part of South Australia.* South Australian Parliamentary Papers 46, Adelaide.
35. Russel H (1898). *Account of an exploration trip led by Mr. H. Russel from Point Virginia to Mount Squires.* Western Australian Parliamentary Papers 45, Perth.
36. Russel H (1899). Expedition to the Barrow, Cavenagh and Warburton Ranges, Western Australia. *Proceedings of the Royal Geographical Society of Australasia (Victoria)* Vol. 17, pp. 64–72.
37. Gara T (1994). "Tackling the back country": Richard Maurice's expeditions in the Great Victoria Desert, 1897–1903. *South Australian Geographical Journal* Vol. 93, pp. 42–60.
38. Gara T (1989). The Aborigines of the Great Victoria Desert: the ethnographic observations of the explorer Richard Maurice. *Journal of the Anthropological Society of South Australia* Vol. 27, pp. 15–47.
39. Maurice RT (1902). A trip to the western interior of South Australia. *Proceedings of the Royal Geographical Society of Australasia, South Australian Branch* Vol. 5, pp. 40–42.
40. Maurice RT (1905). Across the great thirst land. *Wide World Magazine* Vol. 16, pp. 357–364.
41. Murray WR (1902). Mr RT Maurice's expedition north of Fowler's Bay. *Proceedings of the Royal Geographical Society of Australasia, South Australian Branch* Vol. 5, pp. 33–39.
42. Murray WR (1904). *Explorations by RT Maurice — "Fowler's Bay to Rawlinson Ranges" and "Fowler's Bay to Cambridge Gulf."* South Australian Parliamentary Papers 43, Adelaide.
43. Wells LA (1932). *Diary of exploration from Ooldea to Laverton 1932.* PRG 339. Mortlock Library of South Australiana, Adelaide.

44. Hann FH (1903–1907). Unpublished manuscripts and diaries of Frank Hann. Held by the Battye Library, Perth.
45. Donaldson M and Elliot I (eds) (1995). *Do not yield to despair: the 1895–1908 exploration diaries of Frank Hugh Hann.* Facsimile edn to be published by Hesperian Press, WA.
46. George FR (1905). *Prospecting expedition north of Nullarbor Plain.* South Australian Parliamentary Papers 60, Adelaide.
47. *Abstract of the Journal of the Maurice – SA Government prospecting expedition 1904.* Held in the Maurice collection of the Royal Geographical Society of Australasia, South Australian Branch, Adelaide.
48. East N (1905). Unpublished journal of Neal East made during a prospecting trip from Laverton to the Warburton Ranges and return, June 18 – September 27 1905. Department of Conservation and Land Management, Woodvale, Perth.
49. Charlwood, N. Copies of original correspondence from the East Brothers sent to the author by Nell Charlwood, daughter of Neal East.
50. Talbot HWB and Clarke E deC (1917). *A geological reconnaissance of the country between Laverton and the South Australian border, including part of the Mount Margaret Goldfield.* Bulletin No 75. Western Australia Geological Survey, Perth.
51. Talbot HWB (1919). Notes on the country to the north and south of the Transcontinental Railway, between Randall's and the 174 1/2 mile peg. *Western Australia Geological Survey Annual Progress Report 1919*, pp. 74–75.
52. Terry M (1932). *Untold Miles.* Selwyn and Blount, London.
53. Paine HL (1931). *Report and journal of surveyor HL Paine, leader of the Warburton Range Expedition 1931.* Western Australia Department of Lands and Surveys, Perth.
54. Gara T. Personal communication.

Chapter 8 RAILWAYS AND MISSIONS

1. Cocks SM (1978). *Report of the Pitjantjatjara Land Rights Working Party of South Australia.* Adelaide.
2. Gara T, Colley S, Brockwell S and Cane S (1988). *The Aboriginal metropolis of Ooldea.* Internal report to the Aboriginal Heritage Branch, South Australian Department of Environment and Planning. ANUTech, Canberra.
3. Bolam AG (1923). *The Trans-Australian Wonderland.* Modern Printing, Melbourne.
4. White SA (1918). *Ooldea via the East-West Railway.* Held in the library of the South Australian Branch of the Royal Geographical Society of Australasia, Adelaide.
5. Brockwell S, Gara T, Colley S and Cane S (1989). The history of Ooldea Soak and Mission. *Australian Archaeology* Vol. 28, pp. 55–76.
6. Mattingley C and Hampton K (eds) (1992). Ooldea and Yalata (Ch. 29). In *Survival in our own land.* Hodder and Stoughton, Sydney, pp. 235–246.
7. Brady M (1986). Leaving the spinifex: the impact of rations, missions, and the atomic tests on the southern Pitjantjatjara. *Records of the South Australian Museum* Vol. 20, pp. 35–45.
8. Colley S, Brockwell S, Gara T and Cane S (1989). The archaeology of Daisy Bates' campsite at Ooldea, South Australia. *Australian Archaeology* Vol. 28, pp. 79–91.
9. White I (ed.) (1985). Introduction. In Bates D, *The native tribes of Western Australia.* National Library of Australia, Canberra.
10. Bates D (1938). *The passing of the Aborigines.* John Murray, London. First Australian edn.
11. Salter E (1971). *Daisy Bates.* Coward, McCann and Geoghegan, New York.
12. Basedow H (not dated). *Report upon the Second Medical Relief Expedition among the Aborigines of South Australia.* Public Records Office (GRG 23/1/87/1921).
13. Hill E (1973). *Kabbarli. A personal memoir of Daisy Bates.* Angus and Robertson, Sydney.
14. Turner V (1950). *Ooldea.* S John Bacon, Melbourne.
15. Berndt R and C (1951). *From black to white in South Australia.* Cheshire, Melbourne.
16. Berndt R and Johnston T (1942). Death, burial and associated ritual at Ooldea. *Oceania* Vol. 12, pp. 189–208.
17. Berndt R and C (1942). Preliminary report of fieldwork in the Ooldea region, western South Australia. *Oceania* Vol. 12, pp. 305–330.
18. Berndt R and C (1942). Acculturation and native policy. *Oceania* Vol. 13, pp. 52–70.
19. Berndt R and C (1942). Social organisations and institutions. *Oceania* Vol. 13, pp. 144–169.
20. Hampel NA (1977). *Yalata Lutheran Mission 1952–1977.* Yalata Lutheran Mission, SA.
21. Brady M and Palmer K (1988). Dependency and Assertiveness. Three waves of Christianity among Pitjantjatjara people of Ooldea and Yalata (Ch. 16). In Swain T and Rose D (eds), *Aboriginal Australians and Christian Missions — ethnographic and historical studies.* The Australian Association for the Study of Religions, pp. 236–249.
22. Palmer K (1982). *Grey earth and clean sand. The Yalata Aborigines and their homelands.* Western Desert Project, Department of Psychiatry, Flinders University of South Australia, Adelaide.
23. Palmer K (1990). Government policy and Aboriginal aspirations: self management at Yalata. In Tonkinson R and Berndt C (eds), *Going it alone. Prospects for Aboriginal autonomy.* Aboriginal Studies Press, Canberra.
24. Cane S (1990). *A proposal for joint management of the Unnamed Conservation Park. A report to Maralinga Tjarutja.* National Heritage Studies, ACT.
25. Hadfield B. Personal communication.
26. Cane S (1991). *The spinifex people. Report to Paupiyala Tjarutja and ATSIC, Kalgoorlie, WA.* National Heritage Studies, ACT.
27. Serventy V (1961–1962). The Great Victoria Desert. *Geographical Magazine* Vol. 34, pp. 692–712 (May 1961 – April 1962).
28. Serventy V (1973). The Great Victoria Desert (Ch. 1). In *Desert Walkabout.* William Collins, Australia.
29. Morgan M (1986). *A drop in the bucket. The Mount Margaret Story.* United Aborigines Mission, Box Hill, Victoria.
30. Stanton JE (1988). Mt Margaret — missionaries and the aftermath (Ch. 20). In Swain T and Rose D (eds), *Aboriginal*

Australians and Christian missions — ethnographic and historical studies. The Australian Association for the Study of Religions, pp. 292–307.
31. Douglas W. Personal communication.
32. Glass A (1993). *Into another world. A glimpse of the culture of the Ngaanyatjarra people of Central Australia.* Institute for Aboriginal Development, Alice Springs.
33. Green N (1990). *Desert School*. Fremantle Arts Centre Press, WA.
34. Ngaanyatjarra Council (1993). *Karilywarra Community living area land needs: An excision from the Gibson Desert Nature Reserve.* Ngaanyatjarra Council (Aboriginal Corporation), Alice Springs.
35. Edwards WH (1992). Patterns of Aboriginal residence in the north-west of South Australia. *Journal of the Anthropological Society of South Australia* Vol. 30, pp. 2–32.
36. Mattingley C and Hampton K (eds) (1992). Ernabella (Ch. 32). In *Survival in our own land*. Hodder and Stoughton, Sydney, pp. 255–261.

Chapter 9 Rockets and Atomic Tests
1. Milliken R (1986). *No conceivable injury*. Penguin Books, Melbourne.
2. Copley PB (1992). Land Use History. In Copley PB and Kemper CM (eds), *A biological survey of the Yellabinna region South Australia.* National Parks and Wildlife Service and Department of Environment and Land Management, Adelaide, pp. 43–47.
3. Mattingley C and Hampton K (eds) (1992). Atom bombs before Aborigines: Maralinga (Ch. 11). In *Survival in our own land.* Hodder and Stoughton, Sydney.
4. Symonds JL (1985). *A history of British atomic tests in Australia.* AGPS, Canberra.
5. Major RB (1985). *British nuclear weapons tests in Australia. A summary.* Report Book 84/37. Department of Mines and Energy SA.
6. O'Shea D (1991). Maralinga — righting the wrongs. *Habitat Australia*, June issue, pp. 20–23 .
7. McClelland J, Fitch J and Jonas WJA (1985). *The report of the Royal Commission into British nuclear tests in Australia*. AGPS, Canberra.
8. Gara T, Colley S, Brockwell S and Cane S (1988). *The Aboriginal metropolis of Ooldea*. Internal report to the Aboriginal Heritage Branch, Department of Environment and Planning, Adelaide. ANUTech, Canberra.
9. Cane S (1990). *A proposal for joint management of the Unnamed Conservation Park. A report to Maralinga Tjarutja*. National Heritage Studies, ACT.
10. Technical Assessment Group, Department of Primary Industries and Energy (1990). *Rehabilitation of former nuclear test sites in Australia*. AGPS, Canberra.
11. Pearce N (1968). *Final report on residual radioactive contamination of the Maralinga Range and the Emu site.* United Kingdom Atomic Energy Authority, Atomic Weapons Research Establishment Report, 0–16/68 (unpublished).
12. Australian Ionising Radiation Advisory Council Report No 4 (1979). *Radiological safety and future land use at the Maralinga atomic weapons test range.* AGPS, Canberra.
13. Australian Ionising Radiation Advisory Council Report No 9 (1983). *British nuclear tests in Australia — a review of operational safety measures and of public after-effects*. AGPS, Canberra.
14. Kerr CB, Bentley K, Kean DW and Robotham R (1984). *Report of the Expert Committee on the review of data on atmospheric fall-out arising from British nuclear tests in Australia*. Commonwealth Government Printer, Canberra.
15. Brady M (1986). Leaving the spinifex: the impact of rations, missions, and the atomic tests on the southern Pitjantjatjara. *Records of the South Australian Museum* Vol. 20, pp. 35–45.
16. Cane S (1991). *The spinifex people. Report to Paupiyala Tjarutja and ATSIC*, Kalgoorlie, Western Australia. National Heritage Studies, ACT.
17. Personal communications from Ushma Scales and Jan Turner, staff anthropologists with the Pitjantjatjara and Ngaanyatjarra Councils, respectively.
18. Russell D (1981). *A survey of diseases that may be related to radiation among Pitjantjatjara on remote reserves*. South Australian Health Commission (unpublished).
19. Donovan ED, Stevenson CE and Ariotti E (1983). *Health of atomic test personnel*. AGPS, Canberra.
20. Darby SC, Kendall GM, Fell TP et al. (1993). Further follow up of mortality and incidence of cancer in men from the United Kingdom who participated in the United Kingdom's atmospheric nuclear weapon tests and experimental programmes. *British Medical Journal* Vol. 307 (11 Dec.), pp. 1530–1535.
21. Weir L (1993). Plan to bury Maralinga soil in pits. *The Advertiser*, Mon., 21 June, p. 3, Adelaide.
22. Satchell T and Hackett C (1993). Maralinga clean-up set to start. *The Advertiser*, Wed., 30 June, p. 3, Adelaide.
23. Kriven S (1993). A price unpaid for a poisoned land. *The Advertiser*, Sat. 3 July, p. 7, Adelaide.
24. Kelton G (1994). SA moves for A-test site return. *The Advertiser*, Mon., 2 May, Adelaide.
25. Austin N (1994). Aborigines hail $13.5 m bomb compo. *The Advertiser*, Sat., 3 December, p. 3, Adelaide.

Chapter 10 The Beadell Tracks
1. Beadell L (1989). *Too long in the bush*. Weldon Publishing, Sydney.
2. Beadell L (1989). *Blast the bush*. Weldon Publishing, Sydney.
3. Beadell L (1989). *Bush bashers*. Weldon Publishing, Sydney.
4. Beadell L (1989). *Still in the bush*. Weldon Publishing, Sydney.
5. Beadell L (1989). *Beating about the bush*. Weldon Publishing, Sydney.
6. Beadell L (1989). *End of an era*. Weldon Publishing, Sydney.
7. Beadell L (1988). *Outback Highways*. Weldon Publishing, Sydney. (*Outback Highways* contains a selection of articles from the above six books (refs 1–6).)

8. A video on Len Beadell's road-making feats, entitled *Too Long in the Bush — Len Beadell looks back*, was released in 1993. It was produced by Camera Directions Pty Ltd and Len and Anne Beadell.
9. Southall I (1962). *Woomera*. Angus and Robertson, Sydney.
10. Morton P (1989). *Fire across the desert*. AGPS, Canberra.
11. Beadell L (1993). Bush christening. *Australian Geographic* Vol. 30 (April–June issue), p. 19.

Chapter 11 CURRENT ABORIGINAL LAND TENURE AND A RETURN TO TRADITIONAL LANDS

1. Palmer K (1990). Government policy and aspirations: self management at Yalata (Ch. 6). In Tonkinson R and Berndt C (eds), *Going it Alone. Prospects for Aboriginal autonomy*. Aboriginal Studies Press, Canberra.
2. Toyne P and Vachon D (1987). *Growing up the country. The Pitjantjatjara struggle for their land*. Penguin Books, Melbourne.
3. Edwards WH (1992). Patterns of Aboriginal residence in the north-west of South Australia. *Journal of the Anthropological Society of South Australia* Vol. 30, pp. 2–32.
4. Edwards WH (1992). Aboriginal land rights (Ch. 10). In Mattingley C and Hampton K (eds), *Survival in our own land*. Hodder and Stoughton, Sydney.
5. Cocks C (1978). *Report of the Pitjantjatjara Land Rights Working Party of South Australia*. Adelaide.
6. Ngaanyatjarra Council (Aboriginal Corporation) (1993). *Karilywarra Community living area land needs: an excision from the Gibson Desert Nature Reserve*. Ngaanyatjarra Council, Alice Springs.
7. Turner J (Staff Anthropologist, Ngaanyatjarra Council). Personal communication.
8. Durack M (Aboriginal Affairs Planning Authority, West Perth). Personal communication.
9. Aboriginal Affairs Planning Authority Annual Report 1993. Commissioner for Aboriginal Planning, West Perth, WA.
10. Palmer K (1982). *Grey earth and clean sand. The Yalata Aborigines and their homelands*. Western Desert Project, Department of Psychiatry, Flinders University of South Australia, Adelaide.
11. Cane S (1990). *A proposal for joint management of the Unnamed Conservation Park. A report to Maralinga Tjarutja*. National Heritage Studies, ACT.
12. Davis PS and Kirke BK (1991). *Kapi Palya. Meeting the water requirements of an Aboriginal community at Oak Valley, Maralinga Lands, South Australia*. South Australian Health Commission, Adelaide.
13. Cane S (1991). *The spinifex people. Report to Paupiyala Tjarutja and ATSIC, Kalgoorlie, WA*. National Heritage Studies, ACT.
14. UNIYA and Australian Catholic Relief (1994). Mabo: What does it mean? *Briefing Paper 1* (Jan issue).
15. Working Party for Aboriginal Reconciliation (1994). *The Native Title Act 1993 — a plain English guide*. Available from the Conservation Council of South Australia, 120 Wakefield Street, Adelaide.
16. Wootten H (1993). Green and black after Mabo. *Habitat Australia* Vol. 21 (Nov. issue), pp. 18–21.

Chapter 12 CONSERVATION OF THE DESERT'S NATURAL AND CULTURAL HERITAGE

1. Burbidge AA, McKenzie NL, Chapman A and Lambert PM (1976). The wildlife of some existing and proposed reserves in the Great Victoria and Gibson Deserts, Western Australia. *Wildlife Research Bulletin Western Australia* Vol. 5, pp. 1–16.
2. McKenzie NL and Burbidge AA (eds) (1979). The wildlife of some existing and proposed nature reserves in the Gibson, Little Sandy and Great Victoria Deserts, Western Australia. *Wildlife Research Bulletin Western Australia* Vol. 8, pp. 1–36.
3. Pearson D (1994). The vegetation and flora of Queen Victoria Spring Nature Reserve, Western Australia. Unpublished MSc thesis. Department of Botany, University of Western Australia.
4. Chapman A (CALM, Kalgoorlie). Personal communication.
5. Keally I (CALM, Kalgoorlie). Personal communication.
6. Far West Region, Department of Environment and Land Management (1993). *Unnamed Conservation Park Management Plan*. Department of Environment and Land Management, SA.
7. Morelli J (1992). *Fire management in the Great Victoria Desert*. Australian National Parks and Wildlife Service and South Australian National Parks and Wildlife Service.
8. Douglas JS, Williams L, Levitzke V and Taverner B (1979). *Survey of the Unnamed Conservation Park using Landsat*. Ecological Survey Unit, South Australian Department for the Environment, Adelaide.
9. Tynan R (1989). *Tallaringa Lands. Land use allocation recommendations. A summary document*. Rangelands Assessment Unit, Outback Management Region, Department of Lands, Adelaide.
10. Crossley L (1994). *Expedition Tallaringa*. Preliminary Report. Australian and New Zealand Scientific Exploration Society, Melbourne.
11. Copley PB and Kemper CM (1992). *A biological survey of the Yellabinna region South Australia*. National Parks and Wildlife Service and Department of Environment and Land Management, Adelaide.
12. Shephard MDS (1994). *The Simpson Desert — natural history and human endeavour*. Reed, Sydney.
13. Robinson AC, Casperson KD and Copley PB (1990). Breeding records of malleefowl *Leiopa ocellata* and scarlet-chested parrots *Neophema splendida* within the Yellabinna Wilderness Area, South Australia. *South Australian Ornithologist* Vol. 31, pp. 8–12.
14. Vickery J (1967). *Far West and Far North-West interior. South Australian Pastoral Board inspection* (19 June to 19 July 1967). Internal report held by the SA Pastoral Board, Adelaide.
15. McLaren N and Toyne P (1988). *Great Victoria Desert and Nullarbor Park World Heritage listing*. A proposal to the South Australian Government from the Australian Conservation Foundation. Held by the library of the Conservation Council of South Australia, 120 Wakefield Street, Adelaide.
16. Interim Wilderness Committee (1990). *Discussion paper for the proposed Wilderness Protection Act*. Department of Environment and Planning, Adelaide.

17. Wilderness Advisory Committee (1993). *Report to the Minister of Environment and Land Management on Kangaroo Island areas recommended for protection under the Wilderness Protection Act, 1992*. Department of Environment and Land Management, Adelaide.
18. Lesslie R (1991). Wilderness survey and evaluation in Australia. *Australian Geographer* Vol. 22, pp. 35–43.
19. Lesslie R and Taylor S (1983). Wilderness in South Australia. *Occasional Paper No. 1*. Centre for Environmental Studies, University of Adelaide, Adelaide.
20. Sultan R (1991). Introduction. A voice in the wilderness? Aboriginal perspectives on conservation. *Habitat Australia* Vol. 19 (June issue), p. 2.
21. *The review into the management of National Parks and Wildlife Act reserves. Final report 1994*. Department of Environment and Natural Resources, Adelaide.
22. Cane S (1990). *A proposal for joint management of the Unnamed Conservation Park. A report to Maralinga Tjarutja*. National Heritage Studies, ACT.
23. Toyne P and Johnson R (1991). Reconciliation or the new dispossession? Aboriginal land rights and nature conservation. *Habitat Australia* Vol. 19 (June issue), pp. 8–10.
24. Kean J, Richardson G and Trueman N (eds) (1988). *Aboriginal role in nature conservation*. Emu Conference, June 7th–9th, 1988 (held at Emu, SA).
25. Cane S (1991). *The spinifex people. Report to Paupiyala Tjarutja and ATSIC, Kalgoorlie, WA*. National Heritage Studies, ACT.
26. Working Party for Aboriginal Reconciliation (1994). *The Native Title Act 1993 — a plain English guide*. Available from the Conservation Council of South Australia, 120 Wakefield Street, Adelaide.
27. Wootten H (1993). Green and black after Mabo. *Habitat Australia* Vol. 21 (Nov. issue), pp. 18–21.
28. Nutting M (1994). Competing interests or common ground? Aboriginal participation in the management of protected areas. *Habitat Australia* Vol. 22 (Feb. issue), pp. 28–37.
29. Information on the Friends of the Great Victoria Desert Parks can be obtained by writing to the Secretary, George Wood, 21 Light Road, Coromandel Valley SA 5051.
30. Vickery J (1970). Report to the Chairman of the Pastoral Board. Internal correspondence of the South Australian Pastoral Board.
31. Stone M, Kwitko G and Wilson T (1991). *Environmental management guidelines for mineral exploration in the conservation reserves of the Far West District of South Australia*. Department of Mines and Energy SA.
32. Hibburt J (1994). *Petroleum exploration and development in South Australia*. Mines and Energy SA.
33. *Petroleum exploration opportunity. Officer Basin* (1992). Department of Mines and Energy SA.
34. *OF-94A-D Officer Basin, Invitations for applications (April 1994)*. Mines and Energy SA.
35. Carr B and Novak V (Department of Minerals and Energy WA). Personal communication.
36. Flint DJ, Gravestock DI and Newton AW (1988). *An assessment of the mineral potential of Maralinga Lands*. Department of Mines and Energy SA.
37. Lindsay J, Leven J and Krieg G (1993). *Central Officer Basin South Australia. Seismic data acquisition 1993*. Australian Geological Survey Organisation AGSO Record 1992/89.
38. Phillips BJ, James AW and Philip GM (1985). The geology and hydrocarbon potential of the north-western Officer Basin. *APEA Journal*, pp. 52–61.
39. Townson WG (1985). The subsurface geology of the western Officer Basin — results of Shell's 1980–1984 petroleum exploration campaign. *APEA Journal*, pp. 34–51.
40. Anon (1993). Barton bedrock drilling. *Mineral Industry Quarterly* Vol. 72 (Dec. issue), p. 5 .
41. Blockley J (Department of Minerals and Energy WA). Personal communication.
42. Fulwood KE and Barwick RE (1990). Mulga Rock uranium deposits, Officer Basin. In Hughes FE (ed.), *Geology of the mineral deposits of Australia and Papua New Guinea*. The Australian Institute of Mining and Metallurgy, Melbourne, pp. 1621–1623.
43. Duncan RK and Willet GC (1990). Mount Weld carbonatite. In Hughes FE (ed.), *Geology of the mineral deposits of Australia and Papua New Guinea*. The Australian Institute of Mining and Metallurgy, Melbourne, pp. 591–597.
44. Stone M, Kwitko G and Wilson T (1991). *Guidelines for field personnel in mineral exploration in the conservation reserves of the Far West District of South Australia*. Department of Mines and Energy SA.
45. Information Series: Mining Act 1978 (1993). *Guidelines for the application of environmental conditions for exploration and mining on conservation reserves and other environmentally sensitive land*. Department of Minerals and Energy (Geological Survey of Western Australia), East Perth, WA.

Index

Page references in **bold** type indicate photographs.